WALK IN WELL-established corporations

OFFICE DESIGN 国际办公新视野

《国际办公新视野》编委会 主编

大连理工大学出版社

CONTENTS

FACE TO FACE

"面对面"办公室

设计师：佘崇霖
设计公司：Ministry of Design
项目地点：新加坡
项目面积：500 ㎡
摄影师：CI&A Photography

项目介绍 PROJECT DESCRIPTION

"面对面"隶属的 apbcOffices 为企业提供顶级精品服务式写字楼，以最前卫的设计和最先进的技术满足企业管理者的挑剔要求。apbcOffices 正在迅速扩大，目前针对不同行业，在新加坡、吉隆坡、北京和上海设有 11 个业务中心。

整体色调以黑色为主。长廊上的灯柱投射变化多样，几条光带或聚或散，像跳动的音符，沿着天花板进入室内。室内的墙面用平行斜板装饰，白色的影画变化多样，将楼层的墙面连为一体，黑底白影，体现了整体的空间感，狭窄的空间顿时显得深远、广阔，为室内创造黑白交错的光影效果。

The whole space is in black style. The lighting of corridor forms variously, gathering or spreading, leading to the interior like jumping notes. Paralleled plates with various portrayals of human and animal silhouettes provide a quirky and playful sensibility. The graphic diversity allowes the wall to be continious and creates black-and-white crisscross shadows in a black backgroud. These features also aid in the defamiliarization in the normative understanding and experience of typical office spaces.

该设计旨在创造一个轻松愉悦的时尚空间，为人们提供面对面的交流平台。整个项目体现出现代、前卫的设计风格。

The project aims to provide a comfortable environment for communication. It reflects modern and avant-garde design style.

一层

01 人行道
02 通往地下室的旋转楼梯
03 有雕塑感的吧台
04 特色墙，上面有展示壁龛和邮箱
05 停车场
06 储存室
07 食品室
08 特色楼梯
09 卫生间

1ST STOREY

01 FIVE FOOTWAY
02 SPIRAL STAIRS TO BASEMENT
03 SCULPTURAL BAR
04 FEATURE WALL WITH DISPLAY NICHES AND MAILBOXES
05 CAR PARK
06 STORE
07 PANTRY
08 FEATURE STAIRS
09 TOILET

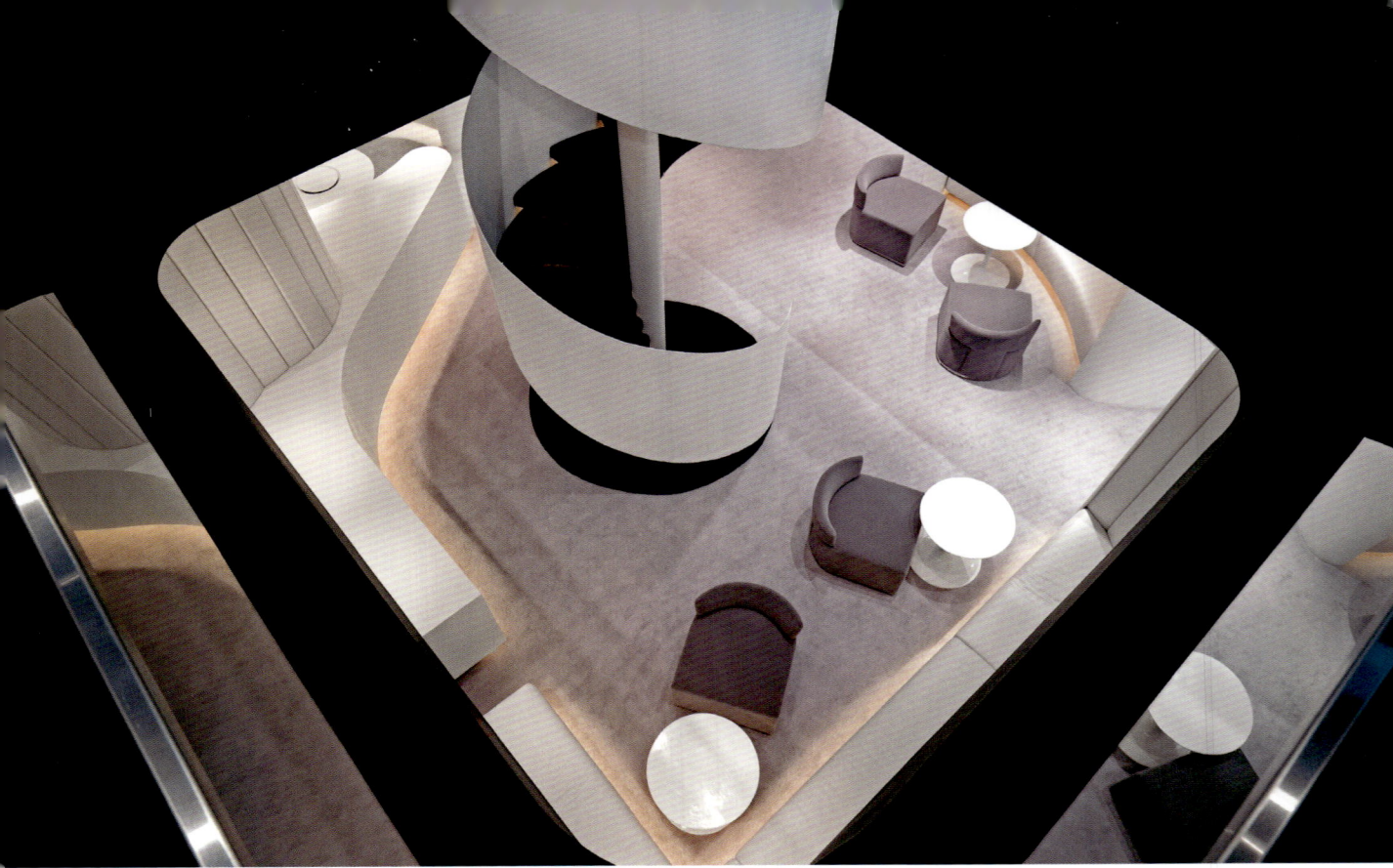

流线型的吧台有助于
互动交流，同时也可
用作特殊场合。吧台
的尽头是一个黑白风
格的旋转楼梯，可直
达底层会议室。底层
与一层截然不同，开
放式的会议室给人舒
适、平等的感觉。三
个大小不一的封闭会
议室内铺着绒毛地毯，
舒适而柔和。

A long sculptural bar formed out of solid surface becomes a magnet for interaction and activity on a daily basis as well as during special functions. At the end of the bar, a rotate stair leading to the ground meeting rooms is also in black-and-white style. The open space in the ground space is completely white different from the above. It allows visitors an experience of comfortable and equal feeling. Three meeting rooms in different sizes have gray downy carpeting and make comfortable sense.

地下室　　　　　BASEMENT

BARWA BANK

Barwa 银行

设计公司：Crea 国际（www.creainternational.com）
项目地点：卡塔尔
项目面积：235 ㎡
摄影师：Jaber Al Azmeh

项目介绍 PROJECT DESCRIPTION

Barwa 银行是一家由卡塔尔中央银行批准设立和管理的银行，隶属于 Barwa 房地产有限公司。该银行提供诸如零售、个人业务、公司业务、房地产金融业务、投资及资产管理的服务，在卡塔尔的经济发展和人民生活中起到了重要的作用。

为强调银行的独特性，自动提款机和分行都采用了同样的设计手法。Barwa 银行的标志——花已经绽放在卡塔尔的大街小巷，传递着为客户提供优质服务的理念。

To strengthen this uniqueness of Barwa Bank all ATM machines and mobile branches have been developed with the same strong design approach. The flower, iconic symbol of Barwa Bank is now blossoming through out Qatar to deliver top quality service to clients.

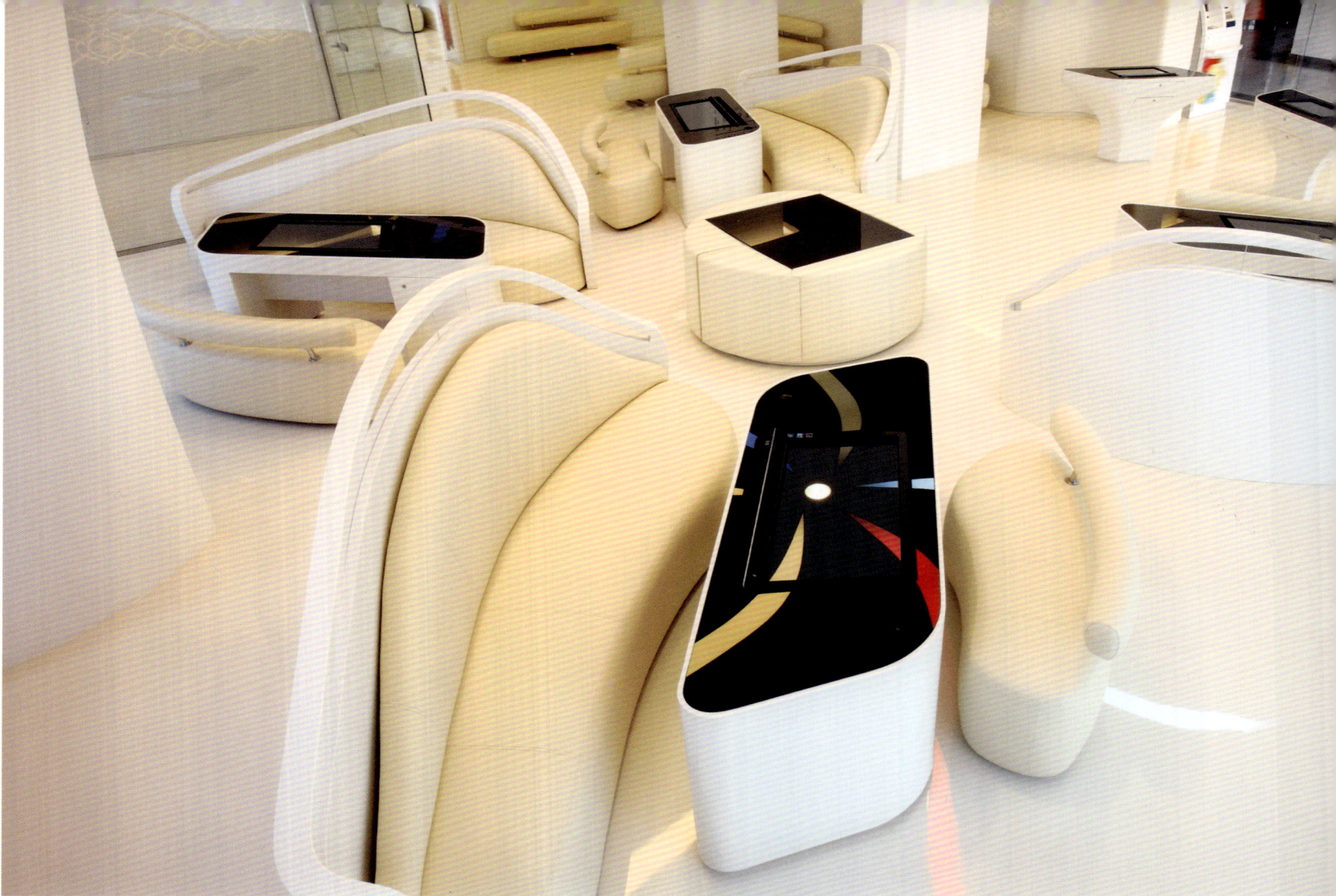

CLIENT 甲方	PROJECT 项目	OBJECT 目标	DATE 日期	SCALE 比例	DRAWING 线图
BAR WA 银行	RETAIL STORE PROFILE DRAWINGS 零售商店线图	GENERAL FURNITURE PLAN 一般家具平面图	25/0110 REVISION 修订	1:75	C1

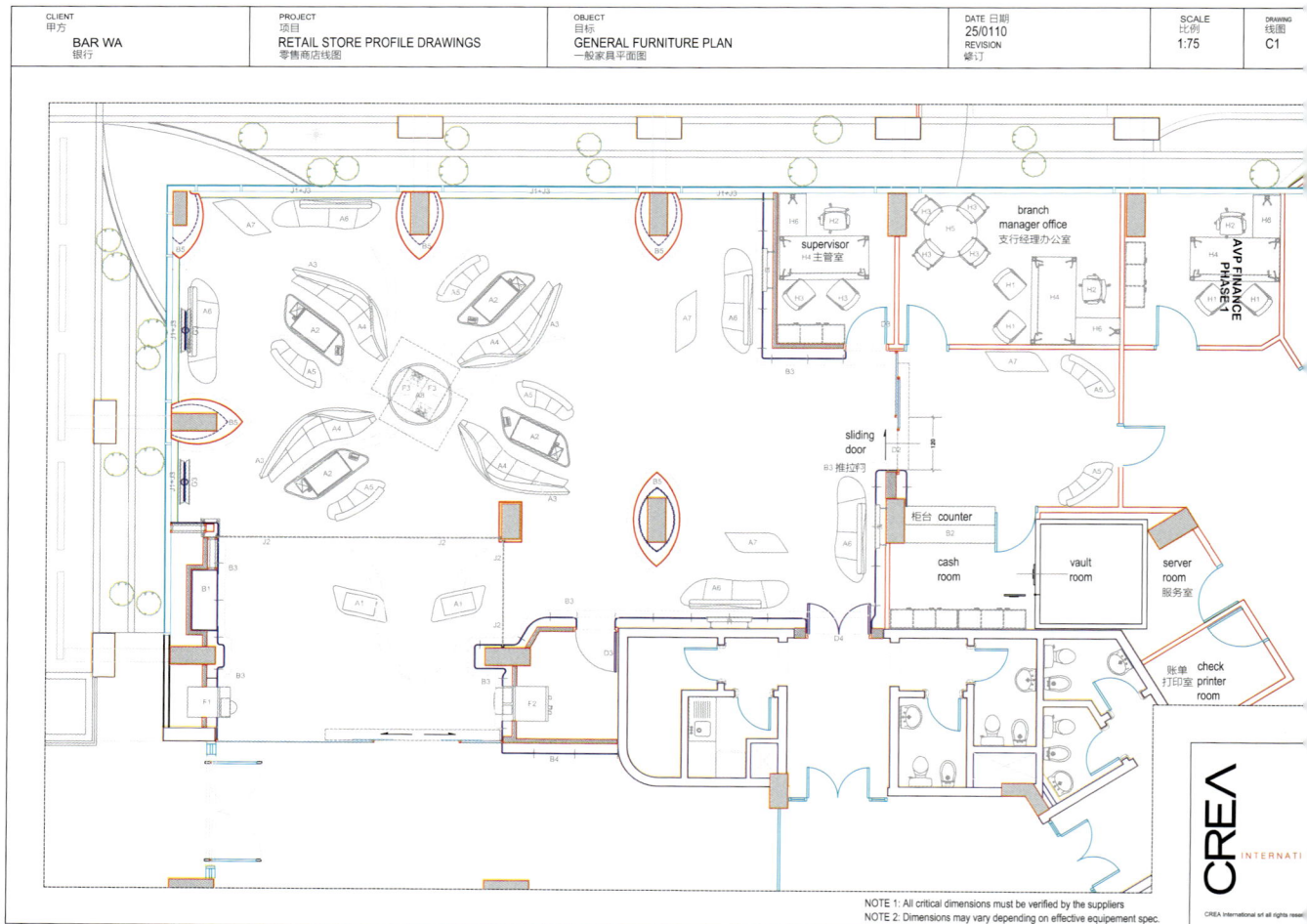

branch manager office 支行经理办公室

supervisor 主管室

AVP FINANCE PHASE 1

sliding door 推拉门

柜台 counter

cash room

vault room

server room 服务室

账单 打印室 check printer room

CREA INTERNATI

NOTE 1: All critical dimensions must be verified by the suppliers
NOTE 2: Dimensions may vary depending on effective equipement spec.

CREA International srl all rights resrv

大厅的中央是几个工作台，像沙漠中珍贵的花朵，呈花瓣状摆开，银行主要的业务活动在这里开展。如珍珠般璀璨的荧光令整个空间充满卡塔尔特色。墙面镶嵌了伊斯兰教风格的蔓藤花纹装饰物。树脂、人造石、法国巴力天花营造出一个流畅、自然的结构，除其中所蕴含的工艺之外，它还为整个银行打造出一个独特、舒适、亲切的环境。

The centre of the branch hosts the precious blossom flower of the desert around which orbit as petals, a series of working stations where the main bank activities sheltering the growth of the country are operated. The mother pearl luminescence dresses the whole space to capture the Qatari colours, while the inlays with Islamic arabesque run along the perimeter walls. Resin, Corian and Barrisol get together to give birth to a fluid natural and morphological shaped architectural box which apart from hiding technology, it imparts to the bank store a unique, cosy and enfolding environmental identity.

"

Barwa 银行是 Crea 国际最精彩的设计项目之一，它引领着未来伊斯兰银行的发展趋势。

The Barwa Bank branch design concept has been one of the most challenging projects that Crea International has ever developed: designing the most progressive Islamic bank of the future.

SUNONE 软件开发公司办公室

设计公司：Department of Architecture Co., Ltd.
项目地点：泰国曼谷
项目面积：1200 ㎡
摄影师：Wison Tungthunya

项目介绍 PROJECT DESCRIPTION

SUNONE 是一款 Sun Microsystems 开发的服务器软件。Sun Microsystems 在行业中被认为是最具创造性的企业之一，它想尝试新的软件方式和定价模式。目前它是唯一一家自己生产电脑和操作系统及其芯片的公司。 Sun 强调唯有"开放"，唯有大家遵守同一标准，才能在互联网上携起手来。

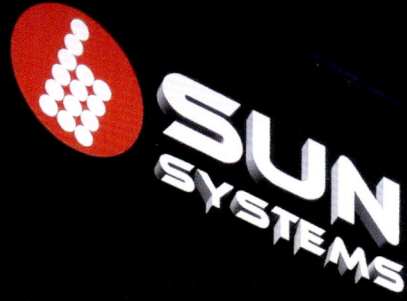

> 该办公室坐落在一片低矮的住宅楼群里，客户希望利用现有的住宅楼条件，为新员工制造良好的工作氛围。既要保留住宅区的安静、私密感，又要创造舒适的办公环境。横格的屏风被运用到该项目中，目的不仅是与室外的住宅环境隔绝，也可以过滤光线，调整室内光照。

Located among a low-rise residential neighborhood, a new location for a software developer – SUNONE presents designers with a great challenge dealing with quality of living for existing residence and working condition for the new comer. In order to preserve serenity of an area, to respect privacy for each residence and to simultaneously create pleasing working environment for staffs, pattern of horizontal screening is introduced to the project.

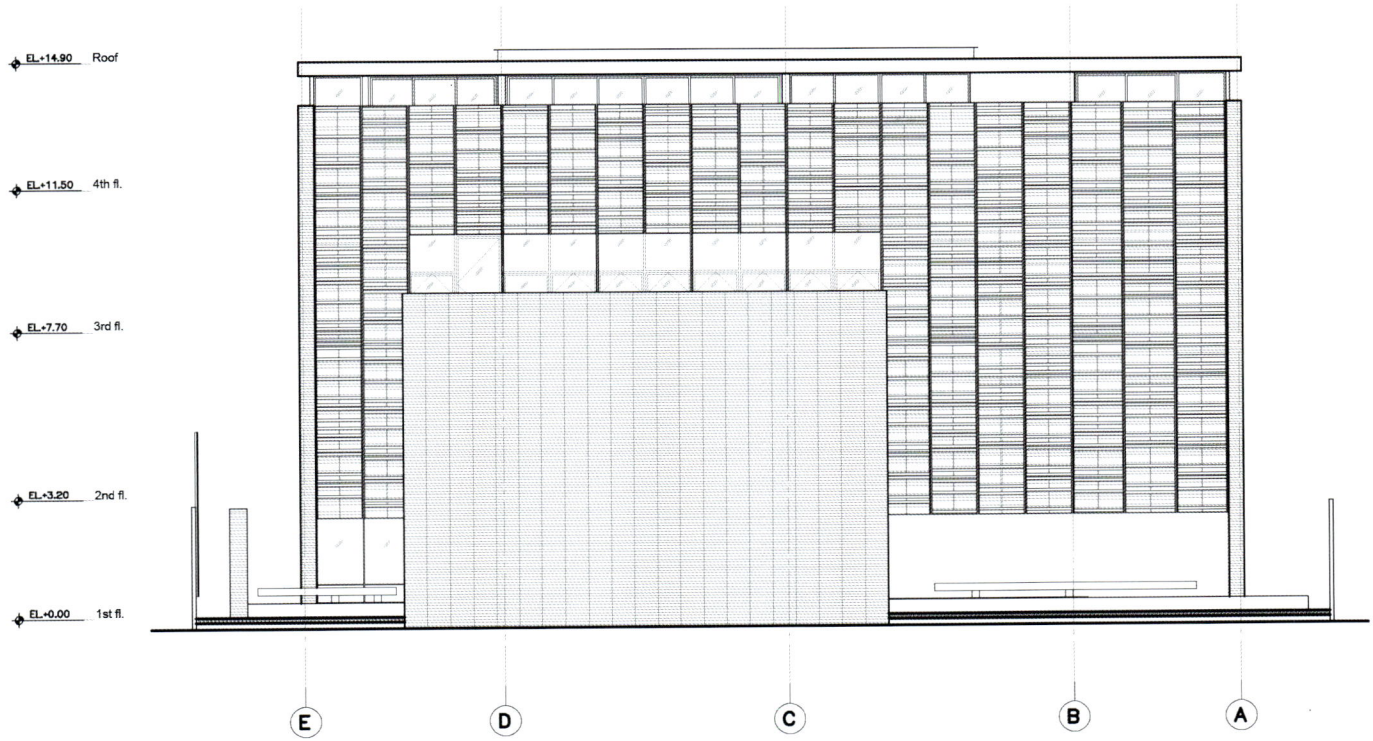

ELEVATION

EL.+14.90	Roof
EL.+11.50	4th fl.
EL.+7.70	3rd fl.
EL.+3.20	2nd fl.
EL.+0.00	1st fl.

E   D   C   B   A

"

整个室内设计格局突出工作环境的灵活和透明。和传统的办公空间不同，SUNONE 需要特殊的长条桌和台灯，也需要许多非正式的休闲椅和吧台作为会议区，社交空间的设计是为了打破IT公司常见的电子交流方式，鼓励面对面交流。设计师通过如此的低调空间，加强了员工的合作与工作的舒适感。

In contrary to the traditional office space, SUNONE provides a range of work settings including bench desks completed with task lights for concentrated work, and informal places such as lounge/bar setting for group meeting and chats. This socially oriented space intends to improve the possibilities for social interaction among staffs rather than previously tended electronic communication.

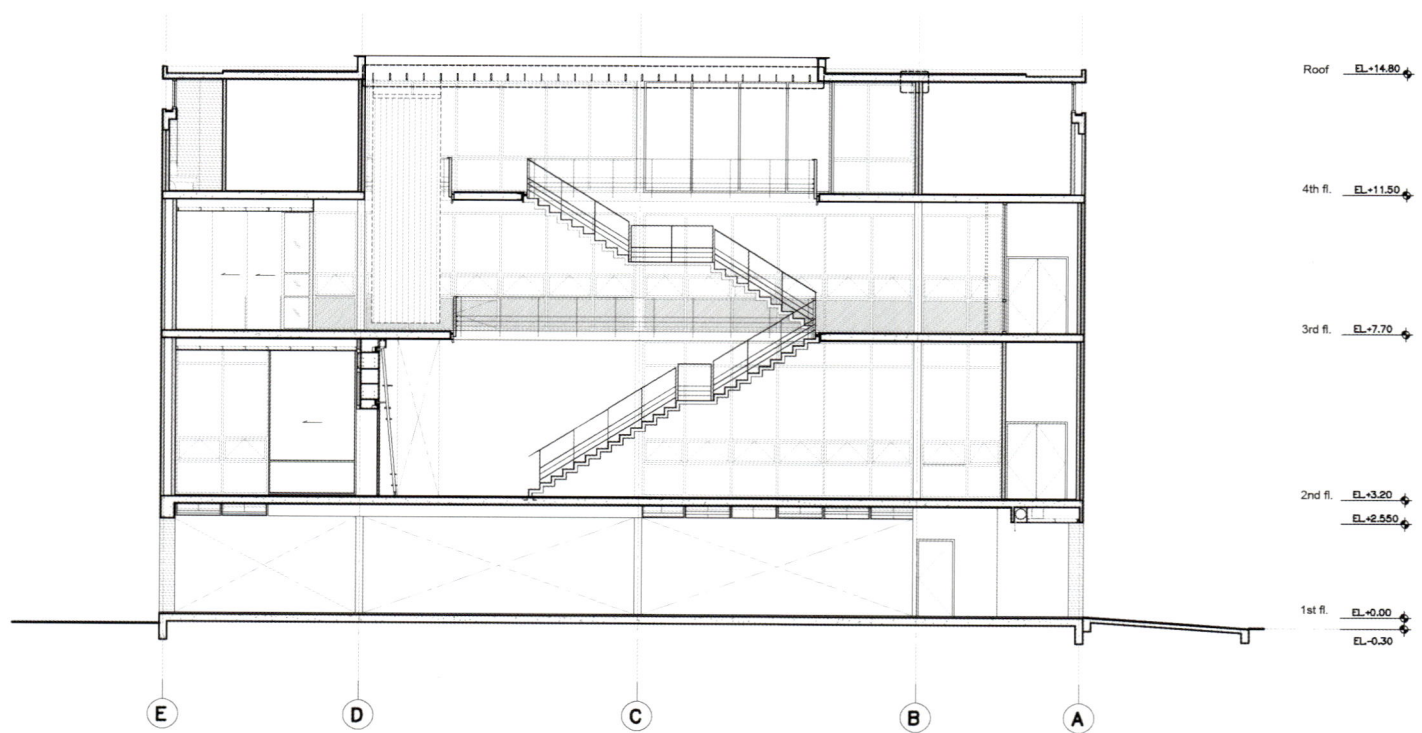

Roof   EL.+14.80

4th fl.   EL.+11.50

3rd fl.   EL.+7.70

2nd fl.   EL.+3.20
            EL.+2.550

1st fl.   EL.+0.00
            EL.-0.30

Ⓔ   Ⓓ   Ⓒ   Ⓑ   Ⓐ

该设计最经典的是屏风，室外的屏风反映了 IT 公司的技术特点，而室内的屏风则直接体现了公司总裁与员工的冒险精神。工业材料如 PVC 彩色条的窗帘巧妙地布置在整个办公室内，充满力量与生机。这种屏风既可以分割空间，又可以起到保护私密的作用，并且办公空间的开放性、透明度都得到了保证。

Truly the most delightful feature of an interior space is the design of interior screen that scattered throughout space. While exterior screen reflects technological character of company's field of business, interior screen, on the opposite, directly reflects adventurous personality of the owner and staff. Industrial material like colorful PVC strip curtain has been cleverly incorporated throughout the building to reflect energy and liveliness of these extreme-sport lovers. Besides operating as space divider and enclosure, this screen, with its transparency property, also functions as an indicator of an area's privacy level.

第二十大道办公室

设计公司：Belzberg 建筑事务所
项目地点：美国
项目面积：465 ㎡
摄影师：Benny Chan/Fotoworks

项目介绍 PROJECT DESCRIPTION

Belzberg 建筑事务所是由一群年轻人组成的团队。办公室的每一个成员都身怀绝技，个性突出。该建筑公司偏好现代、自然的风格，善于在物理环境的限制中运用数码设计，更好地探索建筑结构的创新。

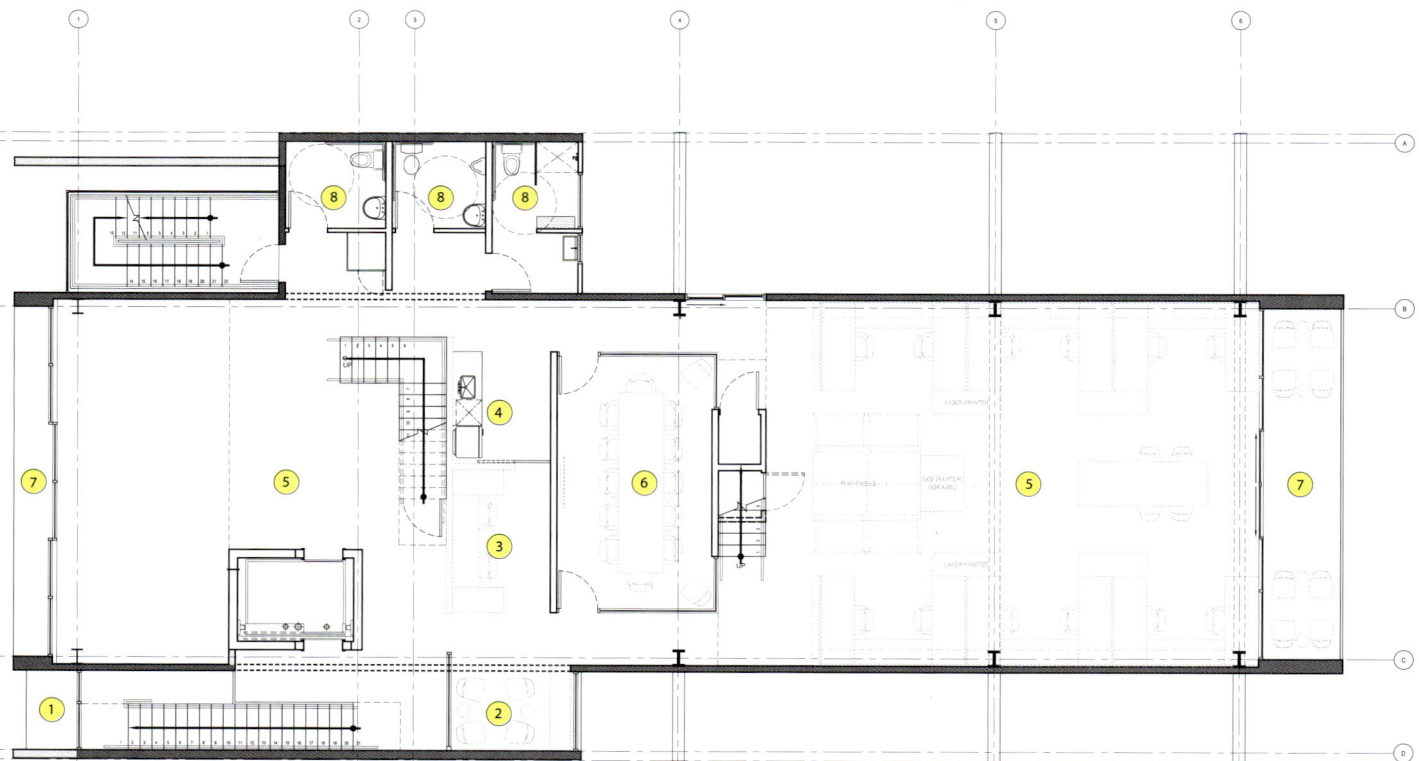

01 通往楼下的入
02 接待处
03 经理室
04 厨房
05 工作室
06 会议室
07 阳台
08 卫生间

01 entry below
02 reception
03 office manager
04 kitchen
05 work studio
06 conference room
07 balcony
08 restroom

"

第二十大道办事处是一项在能效设计方面的探索工程。最初的设计概念是把建筑安置在地面停车场之上，尽量增加绿色空间，优化自然通风和采光。

朝向东面和西面的两个开口使气流进入室内，利于净化空气。建筑的外层包裹着由金属板焊接成的钻石状装饰。自然光线通过建筑两边高 6 米的玻璃窗进入室内，光照面积可超过 75%，减少了人工照明。

The design of the 20th Street Offices was an architectural exploration in efficiency. The initial concept allowed the occupiable space to be lifted above the at-grade parking, maximizing opportunities for open green space, natural ventilation and daylight.

With the open ends oriented to the east and west, the natural flow of air circulates through the tube, maximizing fresh air. The building envelope of the tube element consists of custom designed diamond pattern cladding, fabricated out of sheet metal. The open ends of the tube allow for natural daylight to permeate deep into the space through the 6m tall glazing assembly on either-end, naturally illuminating over 75% of the building and minimizing the need for artificial lighting.

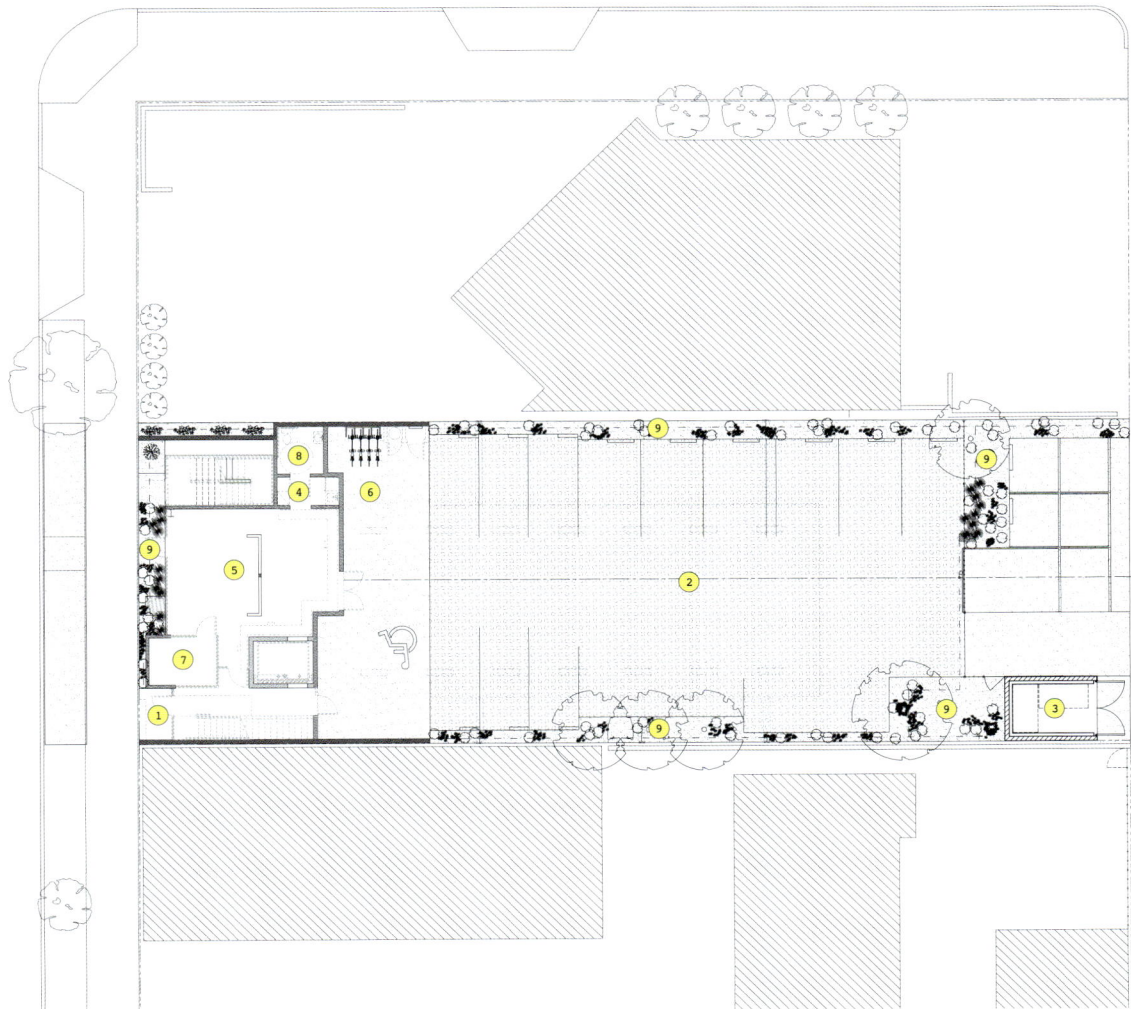

01 入口
02 停车场
03 回收间
04 厨房
05 工作室
06 自行车存放处
07 办公室
08 卫生间
09 绿色空间

01   entry
02   parking
03   recycling
04   kitchen
05   work studio
06   bicycle storage
07   office
08   restroom
09   green space

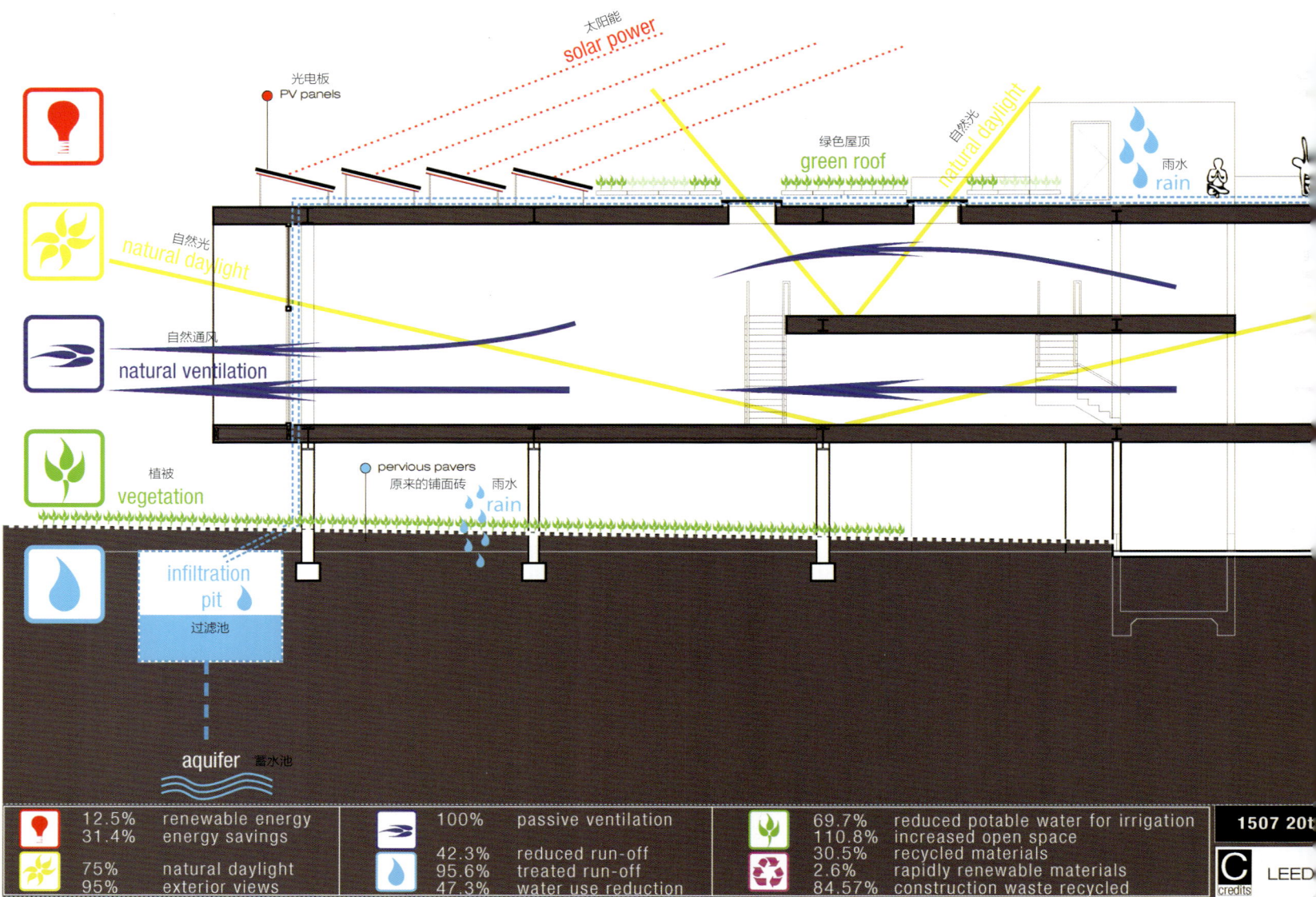

太阳能
solar power

光电板
PV panels

绿色屋顶
green roof

自然光
natural daylight

雨水
rain

自然光
natural daylight

natural daylight

自然通风
natural ventilation

植被
vegetation

pervious pavers
原来的铺面砖

雨水
rain

infiltration
pit
过滤池

aquifer 蓄水池

12.5%	renewable energy		100%	passive ventilation		69.7%	reduced potable water for irrigation
31.4%	energy savings					110.8%	increased open space
			42.3%	reduced run-off		30.5%	recycled materials
75%	natural daylight		95.6%	treated run-off		2.6%	rapidly renewable materials
95%	exterior views		47.3%	water use reduction		84.57%	construction waste recycled

1507 20t

C credits   LEED

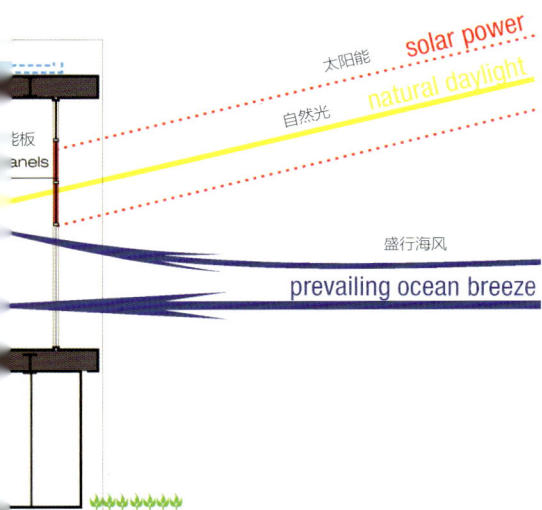

solar power
太阳能
natural daylight
自然光
盛行海风
prevailing ocean breeze

二人屋顶
piable roof

飞板
anels

eet • santa monica • california 90404 | 09

D **45** pts SS WE EA MR EQ ID

"

此外，大楼设计旨在建立一个自然的被动能源系统，为工作人员提供一个高效和清新的工作环境。内部空间分为许多不同功能的空间，用于讨论、产品展示，甚至可以来一堂瑜伽课程。这个设计是在努力创造一种生活方式、一种办公室文化、一种社区与周围环境和谐共生的理念。

The design of the building to primarily function on natural and passive systems makes for a highly efficient and refreshing environment. Broken up into different multifunctional spaces, the offices is for discussions and events, presentations, or even a yoga class. The 20th Street Offices strive to create a lifestyle, an office culture and a connection to the community synonymous with its environmentally conscious informed design.

EASTWEST STUDIOS

Eastwest 录音工作室

设计公司：Hughesumbanhowar Architects、Phillipe Starck
项目地点：美国洛杉矶
项目面积：1906.5 ㎡
摄影师：David Matheson

项目介绍 PROJECT DESCRIPTION

位于好莱坞心脏的 Eastwest 录音工作室是世界上最早的录音棚，五十年来从弗兰克·辛纳特拉到滚石乐队，该工作室制作了许多世界顶级流行音乐。

设计师拆除了所有原来的室内装饰，露出了原始结构和框架。其中只有 5 个模拟式录音室保存完好。该设计具有一系列历史、视觉、体验的层次感。空间的最外层和最里层的外观和软装是复古风格，中间层的墙体和地板设计新颖独特。来访者进入空间后，可以发现高光、奢艳的材料和亚光、做旧的材料形成了强烈的视觉对比。

The designer stripped all interior elements of the project from the building, revealing the original structure and spirit of the construction. The only thing remained are the 5 analog recording studios which are kept wholly intact. The building is a series of historical/visual/experiental layers: vintage on the exterior, fresh and new on the primary layer and again vintage on the inner layer. Shiny and lush materials are contrasted next to dull and weathered materials making for a rich composition as one moves through the space.

" 翻新的空间内包括厨房、宽阔的会客厅、多间私人休息室、露台、会议室和行政办公室。室内灯光较暗，让人有一种与世隔绝的感觉，适合音乐家不受打扰地在其中长时间进行录音工作。

The updated building program now includes a chef's kitchen, spacious lobby, multiple private artist lounges, exterior sun terrace, conference room and programming/administrative offices. The lighting is kept to a minimum in order to reduce the awareness of exterior time, a preferred environment for musicians and their days long recording sessions.

最终的设计效果给来访者一种视听享受的独特体验，既能展示伟大辉煌的录音历史，又饱含对未来大胆而充满趣味的展望。这里成为了一块艺术家能够接触、交流、激情碰撞的地方，同时也影响着未来几代人欣赏音乐的方式。

The result is a truly unique experience that stimulates both the aural and visual senses, that combines the epic sound recordings of the past with a bold, playful vision of the future. It is a place where artists can meet, mingle, and be inspired, while at the same time shaping the way music is heard for generations to come.

PLAJER & FRANZ

plajer & franz 工作室

设计公司：plajer & franz 设计工作室
项目地点：德国
项目面积：1000 ㎡
摄影师：Ken Schluchtmann/diephotodesigner

项目介绍 PROJECT DESCRIPTION

本案为 plajer & franz 工作室，一个 45 位建筑师、室内与图形设计工作人员的创意基地，坐落在德国柏林最大的工业中心。他们用严谨、创新、注重细节的工作态度和一流的品位创造出许多经典之作，现在该工作室是一个享誉国际的设计团队。

整个办公空间以单色调为主。白色和浅色的墙面配以暗色的木材，营造了一个中性风格的空间。波形的前台和镶有高级皮质材料的曲面墙体，以欢迎的姿态迎接人们的来访。墙面背后是存放材料的储物空间。

The monochrome colour scheme that pervades throughout the office is also clearly distinguishable: white and light coloured surfaces in combination with dark wood create a neutral background. The contoured reception counter and curved wall covered in alcantara form an inviting gesture.

楼上的会议室与办公区之间由无框玻璃墙相隔，上面镶有特制的门框。从局域网络到投影机，所有先进科技装备都隐藏于墙、天花板内或家具后，以突出现代的室内设计与原始的建筑结构之间的对比。

顶层螺旋轨道上的窗帘方便人们更加灵活地使用空间，地毯和软装有效地减少了环境中的噪音。会客室空间较小，长沙发、地毯、装饰物在暖色调中营造了一个舒适的环境。墙上嵌入式鱼缸中色彩斑斓的观赏鱼也起到了舒缓心情的作用。

The conference rooms are only separated from other work areas by frameless glass walls with specially designed door frames. All the latest technologies from w-lan to beamer projection are concealed within the walls, ceilings and fixtures to let the brand new design and the existing architecture come to the fore.

On the top floor, a curtain guided in a spiral rail enables a flexible use of the space. The iridescent blue fish of the built-in freshwater aquarium also have a soothing effect.

无论在吧台、长沙发还是大桌子上，工作讨论或者社交都在一个放松的环境下进行。房间的多媒体设备能将这里变成一个俱乐部或者家庭影院。

Whether it's at the bar, on the couch or at the big table, the professional and social exchange here takes place in a relaxed atmosphere. On special occasions, integrated speakers and media technology can transform the space into a club or home cinema.

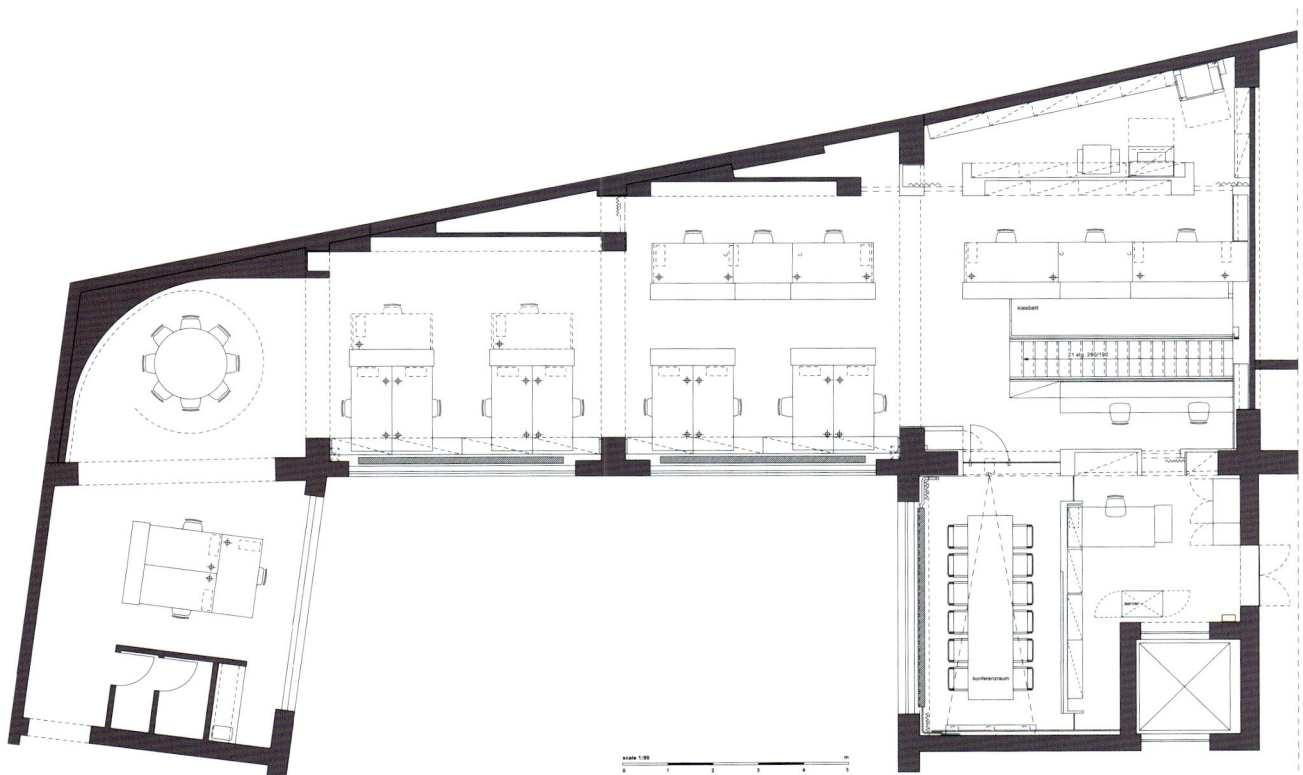

Nike 伦敦总部办公及样品陈列室

设计师：Shaun Fernandes、Simon Jordan、Go Sugimoto
设计公司：Jump Studios
项目地点：英国伦敦
项目面积：1500 ㎡

项目介绍 PROJECT DESCRIPTION

Nike（耐克）被誉为是"近 20 年世界新创建的最成功的消费品公司"。Nike（耐克）运动鞋除了强化高科技运动性能，如今更讲究时尚的外形设计，频频与各国各界潮流达人合作推出联名限量版。

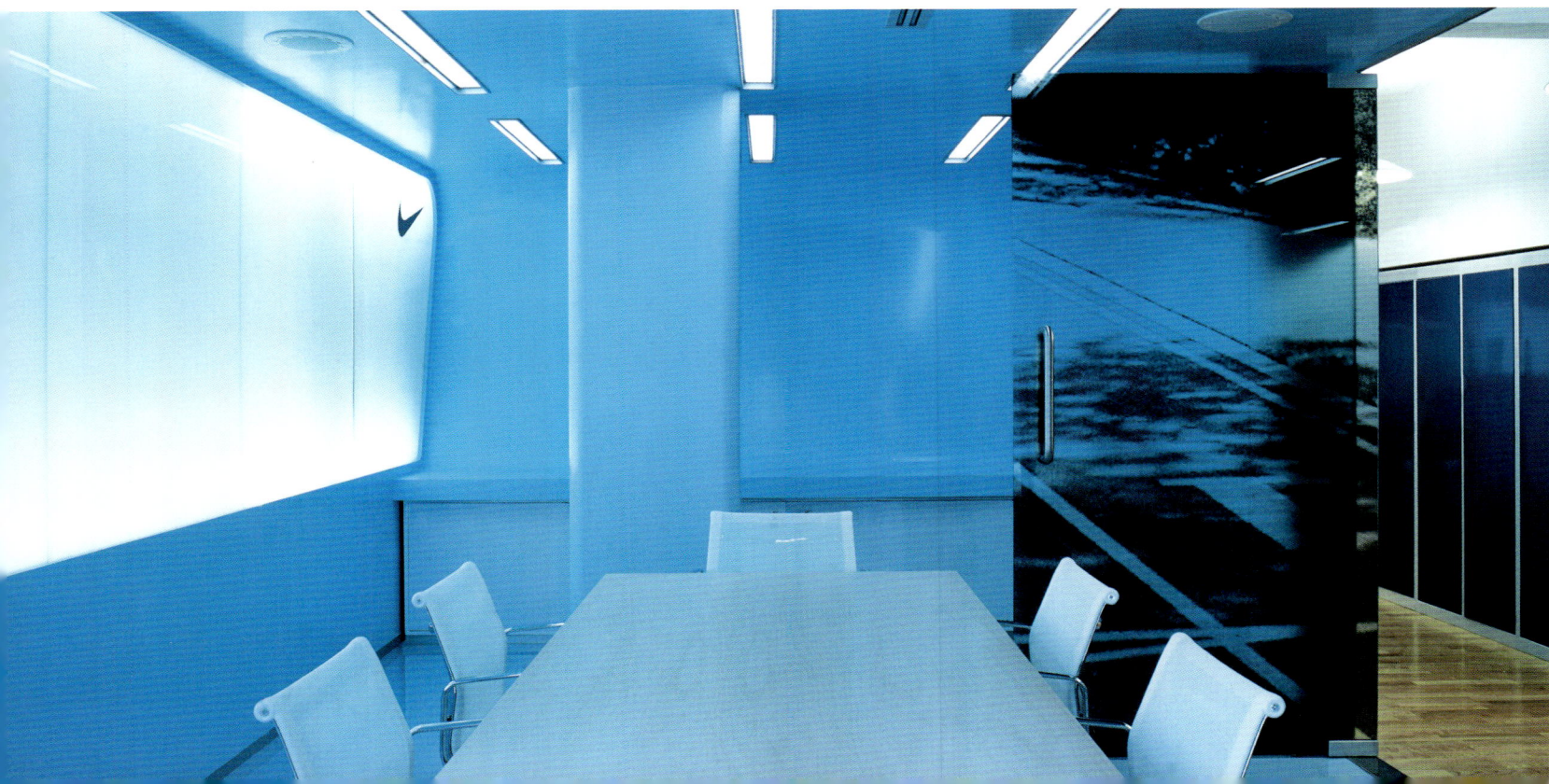

运动系列产品世界中的重量级品牌 Nike（耐克）用伦敦总部样品间重挫竞争对手，他们策略的是让最酷的人穿他们的产品，吸引潮人跟风。旗帜鲜明而积极向上的空间氛围是他们希望通过创意来实现策略的最好证明。

该办公室位于伦敦中心 soho 区域，三层楼 1500 ㎡，囊括了市场部、广告部、市场部、设计部以及样品陈列室。该中心地处全球重要的交通枢纽，人流量很大，所以其设计的灵活性很大，办公区容纳了更多的会面空间——小空间、敞开式咖啡座、封闭式非正式会客室、创意会客厅、正式会议室，这些空间全部配有最先进的技术设备。

The undisputed heavyweight champ of the sportswear world, Nike lands another blow to its envious rivals by grabbing the London HQ showroom. The mission is simple: get the products on people cooler than you, which will hopefully inspire you to follow suit.

The brief has been to create an inspirational and upbeat environment, reflective of Nike's dedication to performance through innovation and its commitment to design as a tool for achieving this.

Located in the heart of soho, central London; situated over three floors and totaling around 1,500m^2, the office houses the marketing, advertising, PR sales and design functions, alongside the product showrooms. As a strategically important global hub, with a constant flow of visiting personnel from around the world, the space has been designed to offer maximum user flexibility, with a high proportion of the working space dedicated to delivering a number of meeting opportunities. These include small "huddle spaces" through to open café areas, enclosed informal meeting rooms, creative meeting spaces and formal conference areas, all of which can accommodate the latest technological working tools.

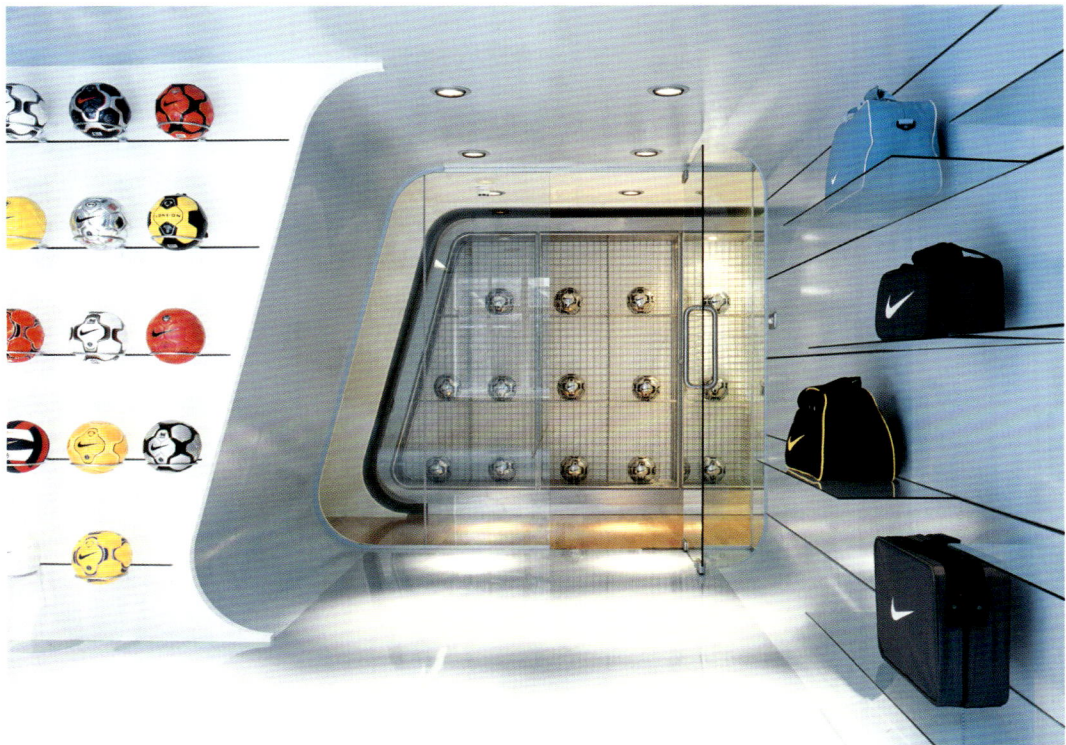

Saatchi & Saatchi 广告公司多伦多办公室

设计师：Inger Bartlett
设计公司：Bartlett & Associates 设计公司
项目地点：澳大利亚多伦多
项目面积：930 ㎡

项目介绍 PROJECT DESCRIPTION

Saatchi & Saatchi 广告公司，隶属全球第三大广告传播集团阳狮集团旗下，总部在纽约。目前已经发展到全球 76 个国家的 140 个分支机构。

办公空间的前台接待区用弧线的网隔从顶上罩住，突出 Saatchi 的设计实力，并分隔出了员工会面区、客户接待区、周会区等地方。金属线鹿头雕塑被粗糙的木条背景衬托得更加显眼。淡黄色的枝杈既缓和了跳跃的现代风格，又同网隔的质地形成对比。

Visitors enter a large reception area defined by curving mesh space frames that reinforce the power of Saatchi and provide room for staff meetings, client receptions and weekly informal socializing that reinforces community. A metal sculpture of a deer head grounds the "Made in Canada" character, further emphasized by rough wooden slats. Dogwood branches soften the urban cityscape outside and provide textural contrast to the mesh screening.

主管区域
**EXECUTIVE
AREA**

普通工作区
GENERAL WORKSPACE

接待区
RECEPTION

咖啡厅
CAFE

消息室
**PITCH
ROOM**

休息室
**BREAKOUT
ROOMS**

会计室
ACCOUNTING

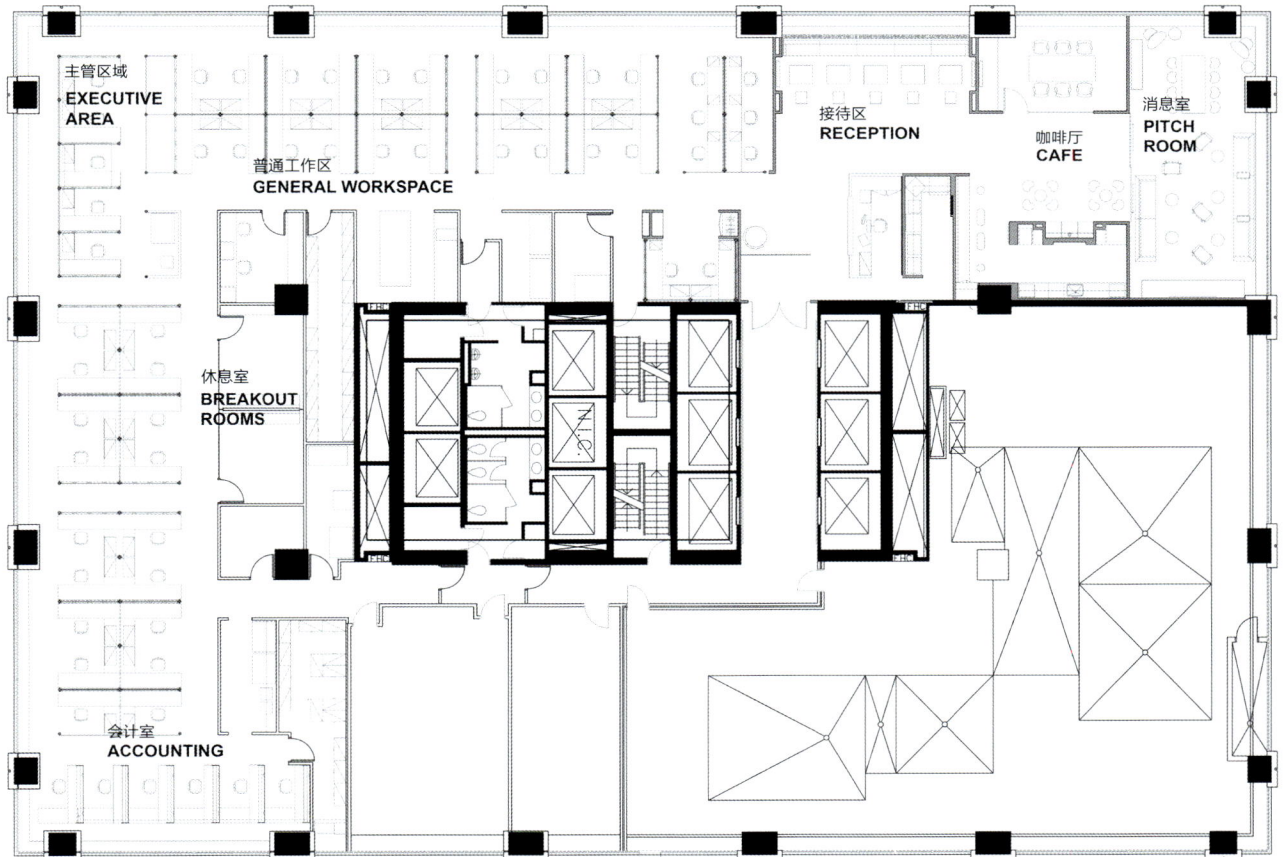

FURNITURE PLAN – SAATCHI　　SAATCHI 家具布置图

该项目中处处充满了象征意义：前台桌椅制造成加拿大荒原的感觉、红色座椅上象征 Saatchi 品牌的心形元素、白色的 Loft 空间顶上使用舞台幕布一样的背景来制造层次感。

Iconic references appear in the integrated exhibition and working space; Eames, Wegner and Saarinen pieces; references to Canadian wilderness; and signature Saatchi elements such as the Lovemark captured in a red Vitra chair. The white loft space provides a stage-like backdrop for layering.

消息室有20个座位，这个地方也是创意团队玩头脑风暴的地方。展览室被设计成了脚手架结构，同时也是定制的工作台架子。

The Pitch Room provides seating for 20 and is also a haven for creative group thinking. The exhibition space is defined by scaffolding structures (associated with architectural creation) that are also the backbone of custom-designed workstations.

这些设计元素将室内设计与家具的设置很好地统一起来。出现在前台的粗胡桃木条和不太正规的脚手架搭配在一起，形成了工作台，同样的材料也贯用在了休息室的墙板、桌子和消息室的门上。各式各样的椅子也使用了胡桃木板材。

Design elements provide unity between the interior and the furniture. The rough walnut slats (in keeping with the informal scaffolding) in Reception reappear as workstation veneers, in wall paneling and tables in the Breakout rooms, and in the Pitch Room doors. Walnut is revisited as moulded plywood in the Eames and Wegner chairs.

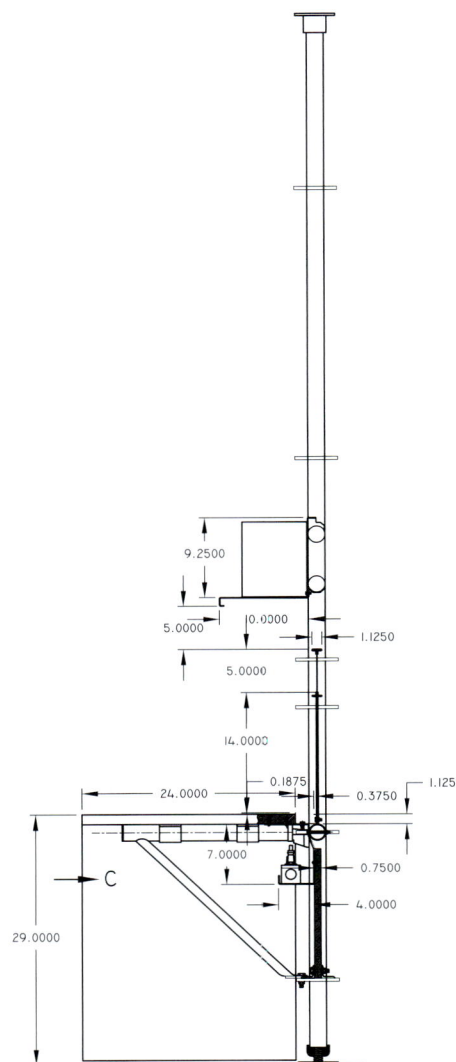

9.2500

5.0000

0.0000

1.1250

5.0000

14.0000

24.0000   0.1875   0.3750   1.125

7.0000

0.7500

C

4.0000

29.0000

C. F. MØller 建筑事务所总部

设计公司：C. F. MØller 建筑事务所
项目地点：丹麦
项目面积：3500 ㎡
摄影师：Julian Weyer、Kontraframe

项目介绍 PROJECT DESCRIPTION

C. F. MØller 建筑事务所是斯堪的纳维亚半岛上最古老和最大的建筑事务所。事务所的工作涉及广泛的专业知识，简单、明确、谦逊和理想这四个信念，从公司 1924 年成立以来，就成为指导事务所工作的原则，并且不断得到重新定义和解释，来满足特定地点项目的国际化趋势和区域性特征。

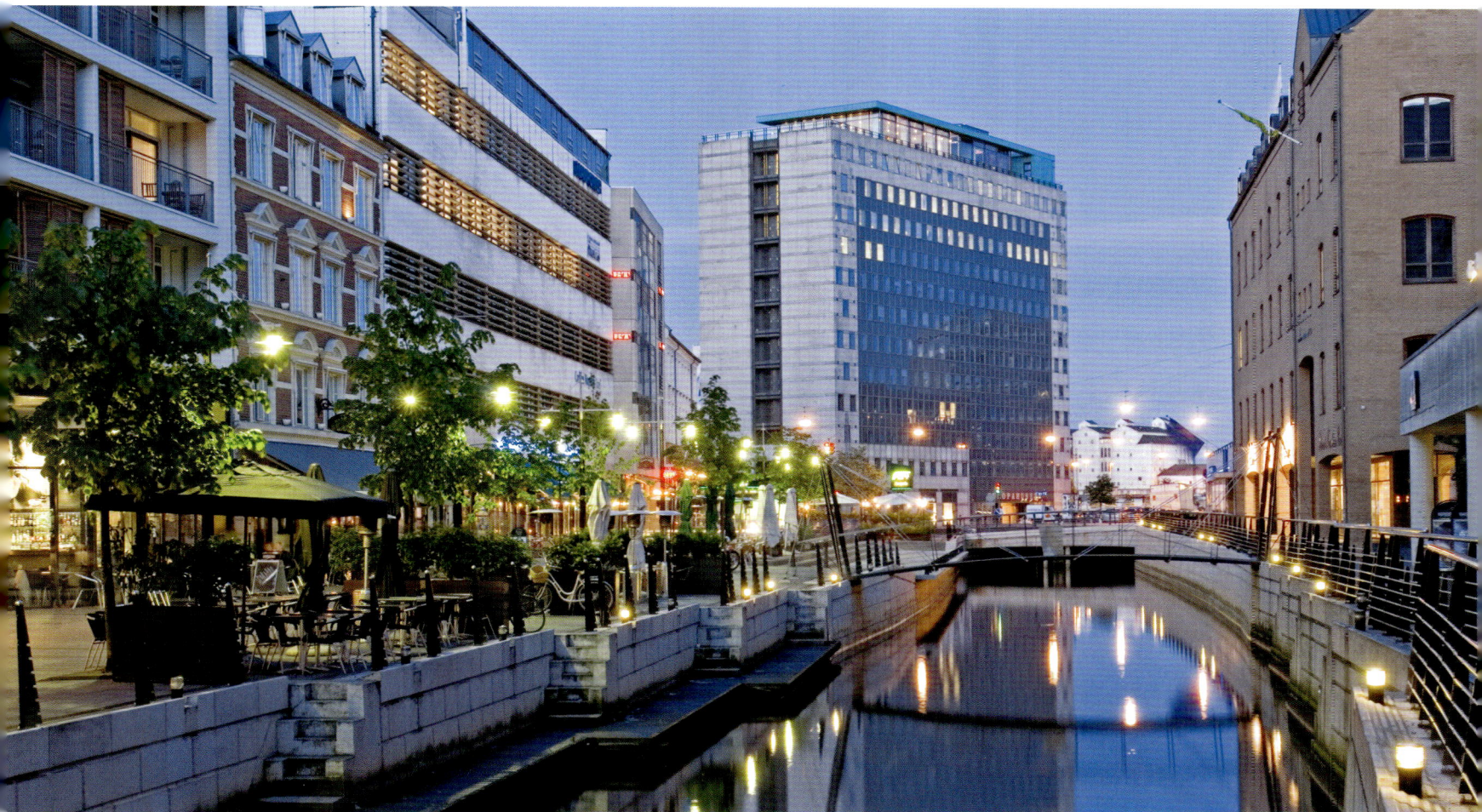

C. F. Møller 建筑事务所的阁楼层有很好的采景，可以看到大街与大厦其他 12 层楼。阁楼上的天台是观察城市河流、教堂和海岸的绝佳位置。

The floor level of the penthouse has been raised up to provide visual contact with the street, twelve storeys below. In this way, the terrace becomes a "presentation tray" for some of the finest urban features in Aarhus: the Aarhus River, the Cathedral and the old coastline.

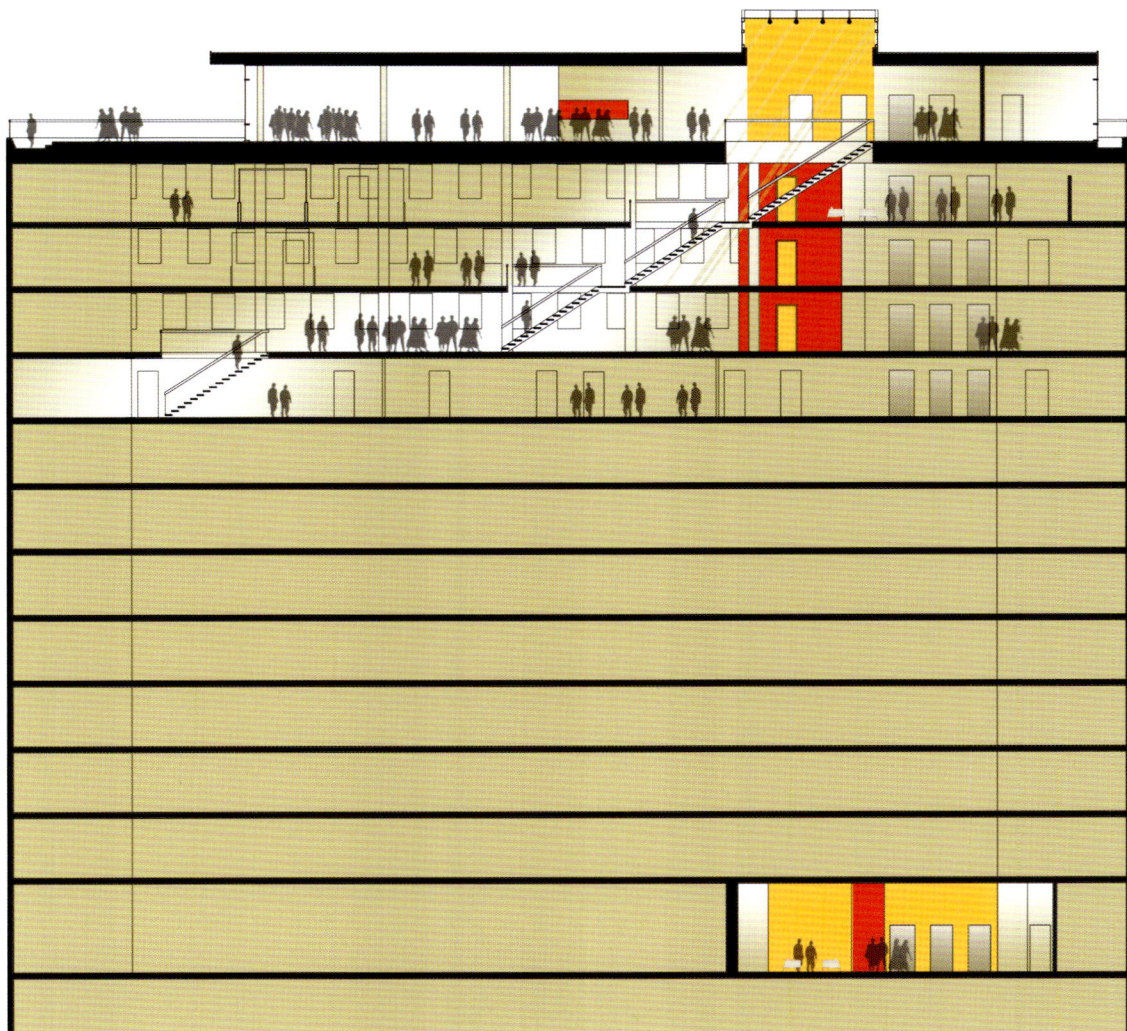

> C. F. Møller 建筑事务所总部坐落在城市中心的一座旧大楼的最高三层。建筑公司将这里重新布局，建造了一个新的阁楼层餐厅、会议室和几个宽阔的屋顶天台，可以容纳 120 名员工办公，里面的结构全部是镀铜玻璃。

C. F. Møller Architects installed its new head offices for approx. 120 co-workers in the top three floors of the building, remodelling the floor plans and constructing a new penthouse storey canteen, meeting facilities and several generous roof terraces.

該办公室包括三层的连续开放式楼层布局，由一部造型楼梯从建筑的中空处贯穿上下，会议室由可以当写字板的玻璃墙隔断。

The design studio occupies 3 coherent open-plan storeys, linked by a single long, sculptural stairway that passes through all the storeys, cutting through the floor slabs. Meeting rooms with "sketchable" glass walls subdivide the plans.

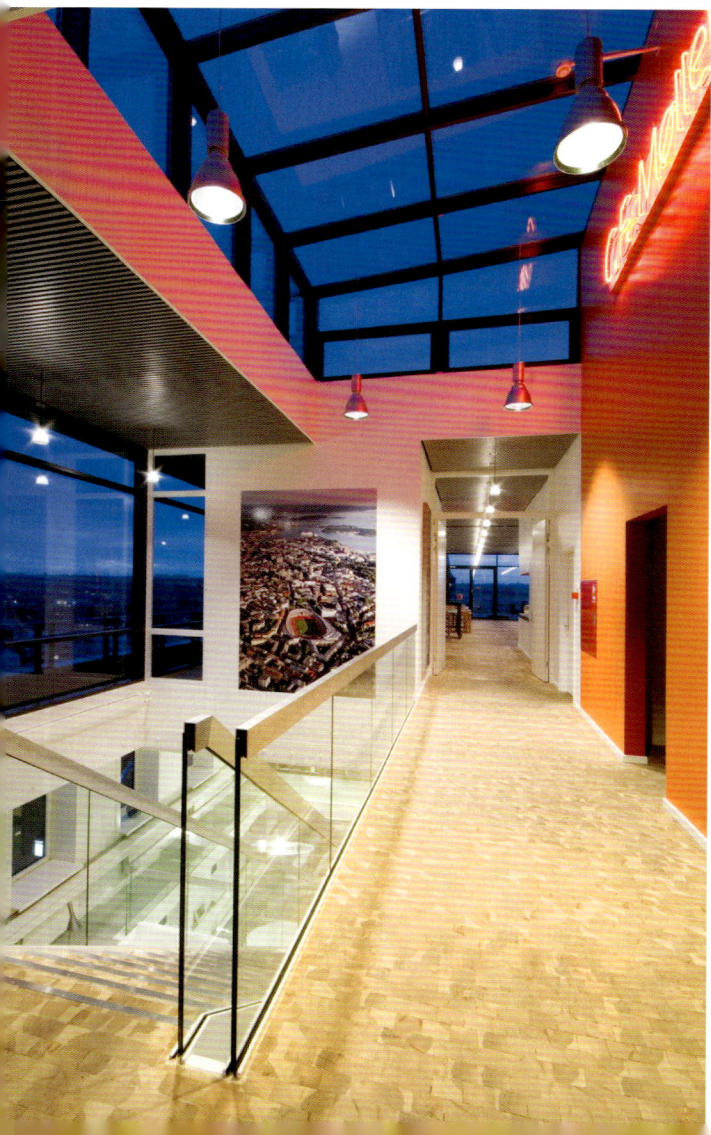

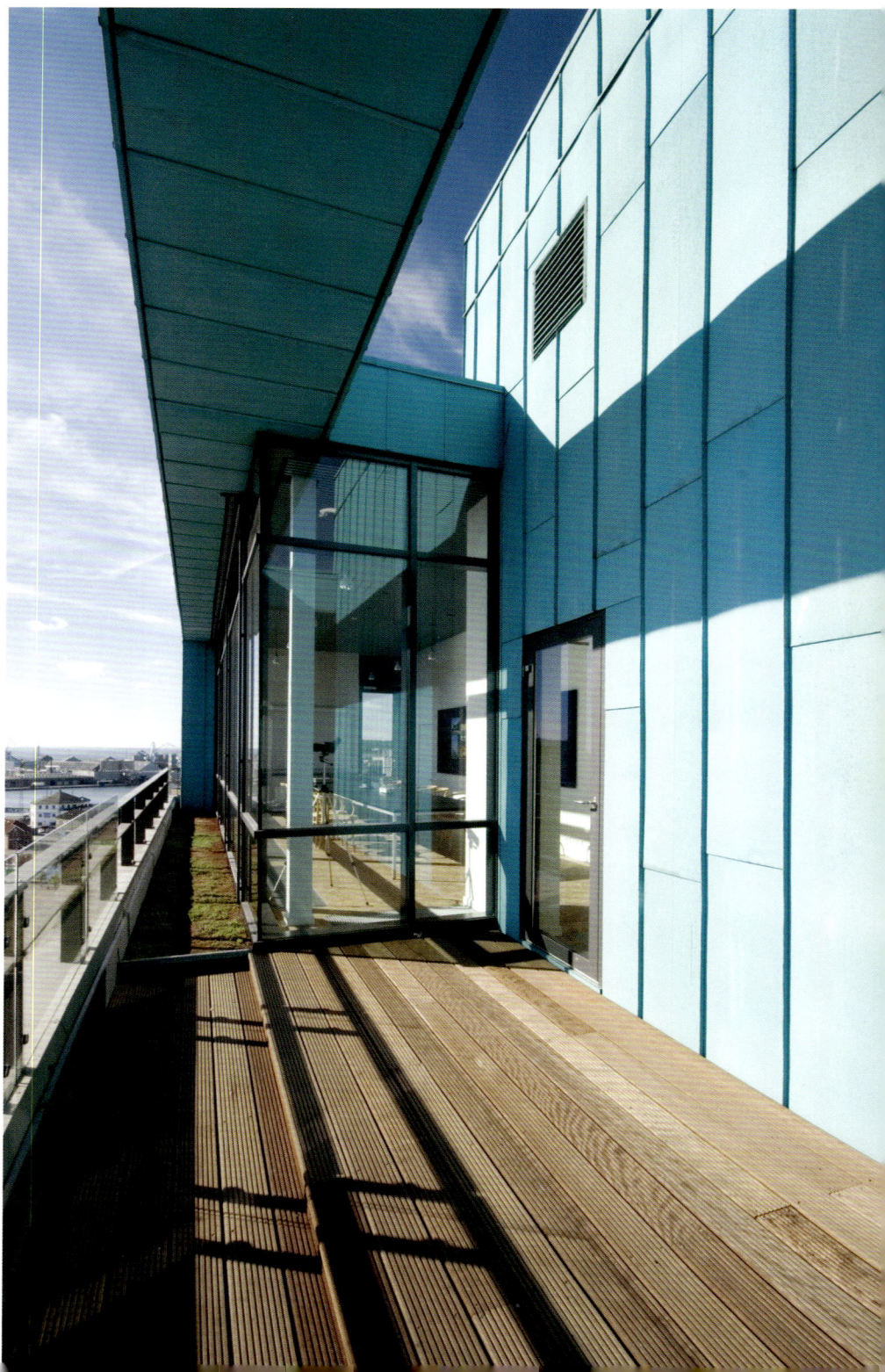

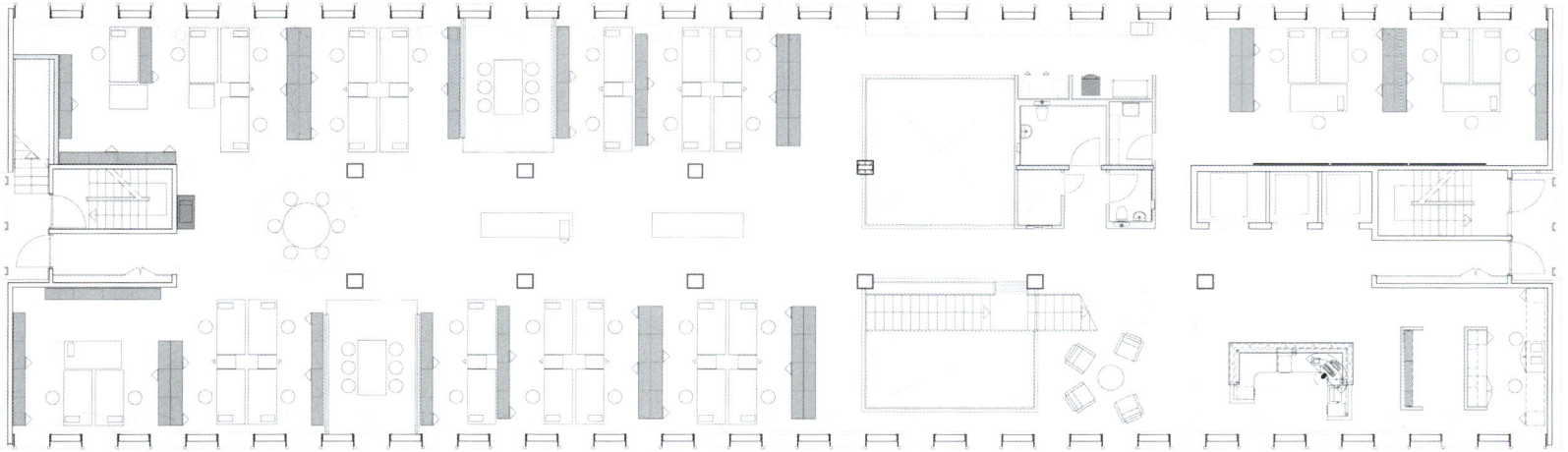

ENDEAVOR

Endeavor 精英经纪公司

设计公司：Neil M. Denari 建筑事务所
项目地点：美国加利福尼亚州
项目面积：651 m²
摄影师：Benny Chan/Fotoworks

项目介绍 PROJECT DESCRIPTION

Endeavor，好莱坞五大经纪公司之一，旗下有大批知名导演、演员和作家。2008 年，一家从事电视业行销的网站进行的独立研究表明，Endeavor 已经把电视界的半数精英揽于帐下。它曾独立制作《越狱》等著名美剧。

办公室依据大楼原来的混凝土结构设计，通道围绕大楼连通各经纪人和助手的办公室。经纪人和助手在相邻区域办公，工作台在门口，助手可直接观察到办公室的情况。

The program and spatial organization is based on the dimensions of the existing concrete building and the circulation itself is entirely predetermined and has to provide access to all offices between the agents and their assistants. All agents have window offices along the perimeter with their assistants situated within the adjacent open office space.

大厅、会议室的天花板和墙面设计新颖独特。工作区的开放布局使更多的光线进入室内。办公楼分为四个独立区域，每个区域有不同的色调：品红色、蓝色、橙色和绿色，各色的壁纸上都有不同的图案。除了以颜色来区分开放的办公区外，洗手间、厨房、电子壁橱、门房等上也有识别的标记。

放映室不仅包括电影放映区，还包括一个大会客室、厨房。布局面向街道，使空间更为开放。这样设计的目的之一是打破周围的平面构造、增强空间感，其次是能让路人看到办公室内的地板和天花板布局，满足人们的好奇心。

The more expressive aspects of the project can be found in ceiling/wall deformations surrounding the main lobby and conference room areas. The main premise of the open office space is to allow for daylight to reach into the space to enhance the natural lighting conditions. In addition to color coding the open offices, symbols are introduced for certain support programs as washroom, kitchen, electrical closet, janitorial, etc.

The screening room consists not only of a high performance space for viewing films, but also a large pre-function lobby area, a kitchen/bar, and a major façade to the street. The purpose is to create as much spatial depth as possible in the entry sequence so as to break down the flatness of the "storefront" conditions surrounding it. The other is to create a public identity for Endeavor and allow passersby to peek in, through, and around these surfaces to catch fragments of the floor and ceiling surfaces inside.

EEA & Tax 办公室

设计公司：UNStudio
项目地点：荷兰
项目面积：48040 ㎡
摄影师：Christian Richters、Ewout Huibers、Ronald Tilleman

项目介绍 PROJECT DESCRIPTION

荷兰国家税务机关和学生贷款管理中心是荷兰的两个公共机构。荷兰国家税务机关负责推行荷兰的税务政策，制定合理的现代税收制度，执行税收征管。荷兰学生贷款管理中心承担着发展学生资助系统、开展信息运动、扩大"未被充分代表"群体的入学率、排除教育系统结构障碍等几个方面的任务，对荷兰高等教育入学政策的变革起到重要作用。

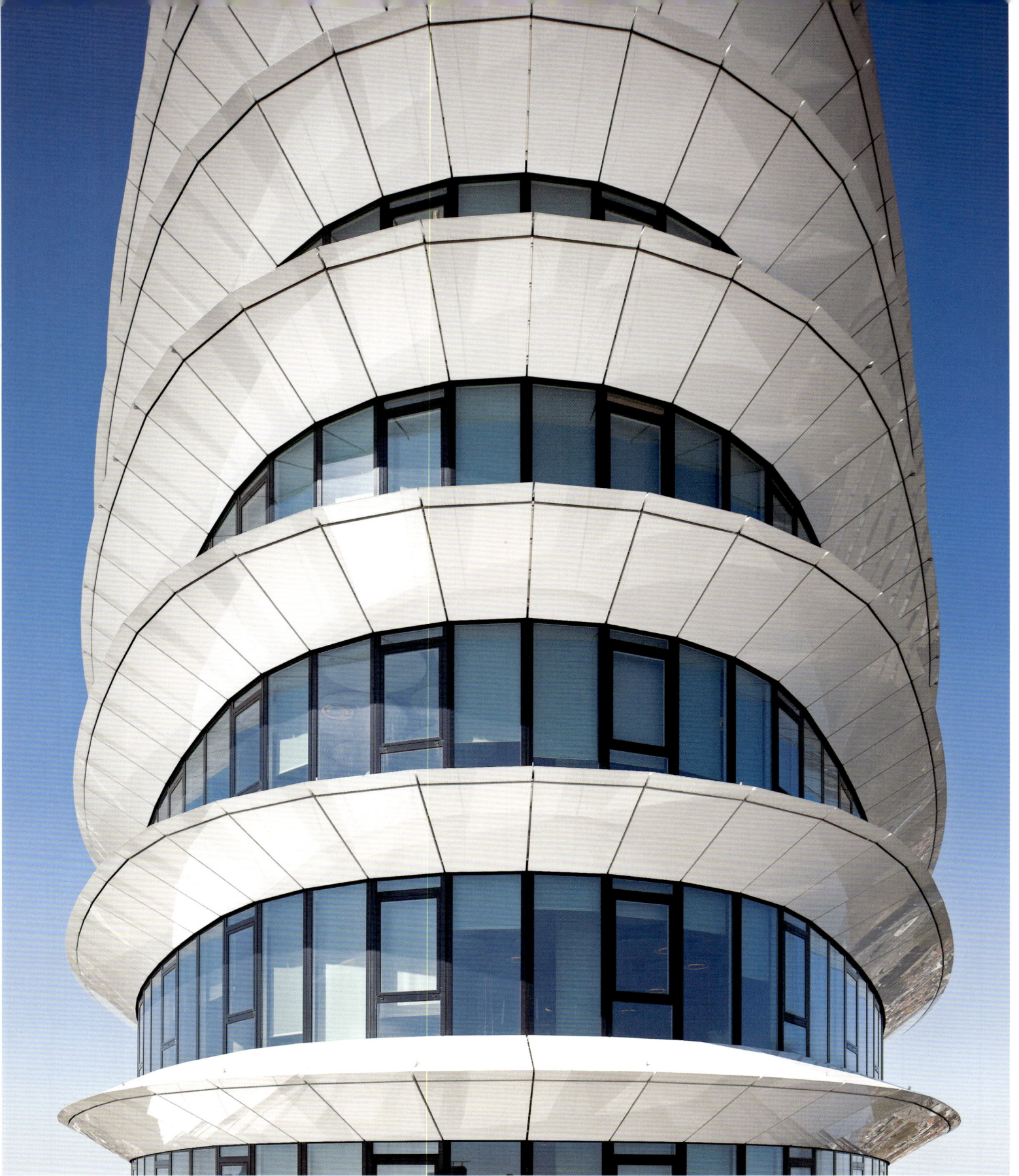

这座表面曲线呈波浪状、不对称的建筑位于一片小森林里，内部是两个公共机构，国家税务机关和学生贷款管理中心。建筑周围有一个大型公共城市花园和具有商业性的多功能展览馆。这个项目的目的是营造一个更加人性化、舒适的环境。它的外观巧妙地掩饰了公立机构的压迫感，打造了一个有机统一、友好现代的建筑。

This undulating curves and asymmetric construction is set amidst small, ancient woodland. The project includes the design, construction and financing of two public institutions; the national tax offices and the student loan administration. The building will be surrounded by a large public city garden with pond and a multifunctional pavilion with commercial functions. The architecture aims to present these institutions with a softer, more human and approachable profile. It deliberately cloaks a commanding public institution in an organic, friendlier and more future-oriented form.

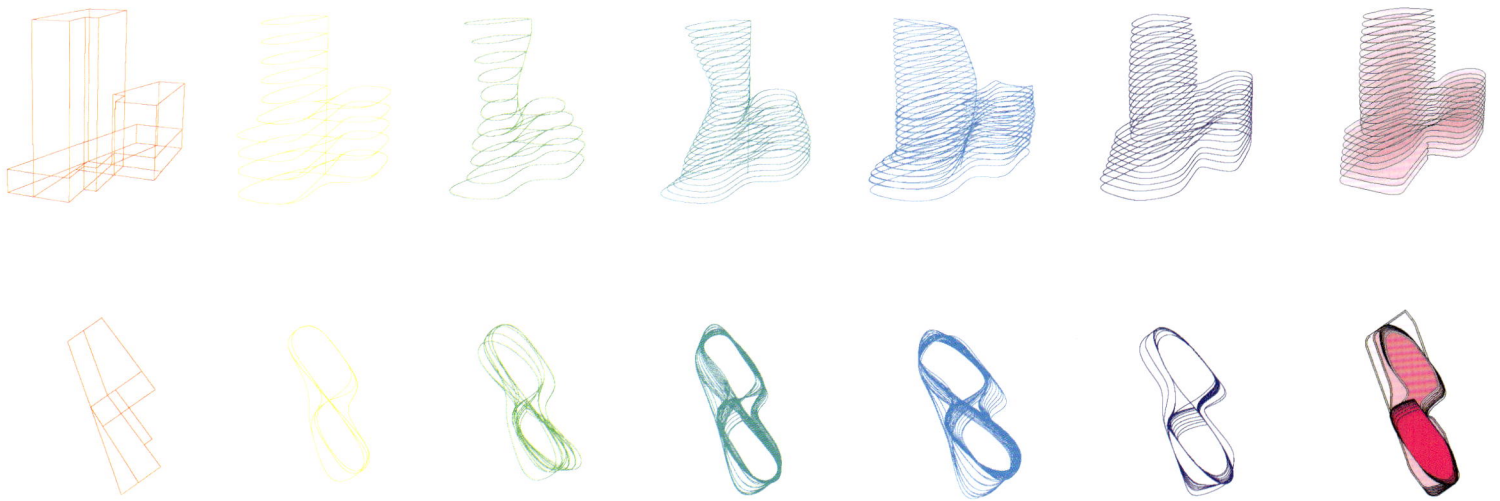

建筑的墙体集遮光、风力控制、光线控制、散热系统功能为一体，另一个特点是混凝土核心活化和地下能源储备装置，这很有效地减少了对外部能源的需求。

多样化的办公环境和有效的节能设计构建了一个超环保的高层建筑，成为荷兰最具有可持续特点的办公大楼之一。

The facade concept integrates shading, wind control, daylight penetration and construction in fin-shaped elements. Another technical feature of the building that contributes to its sustainable character is the combination of concrete core activation and underground long term energy storage. This appreciably reduces the demand for external energy sources.

The inclusion of diverse passive and active environmental and energy efficient solutions has led to a greener approach to tall building which is one of the most sustainable office buildings in the Netherlands.

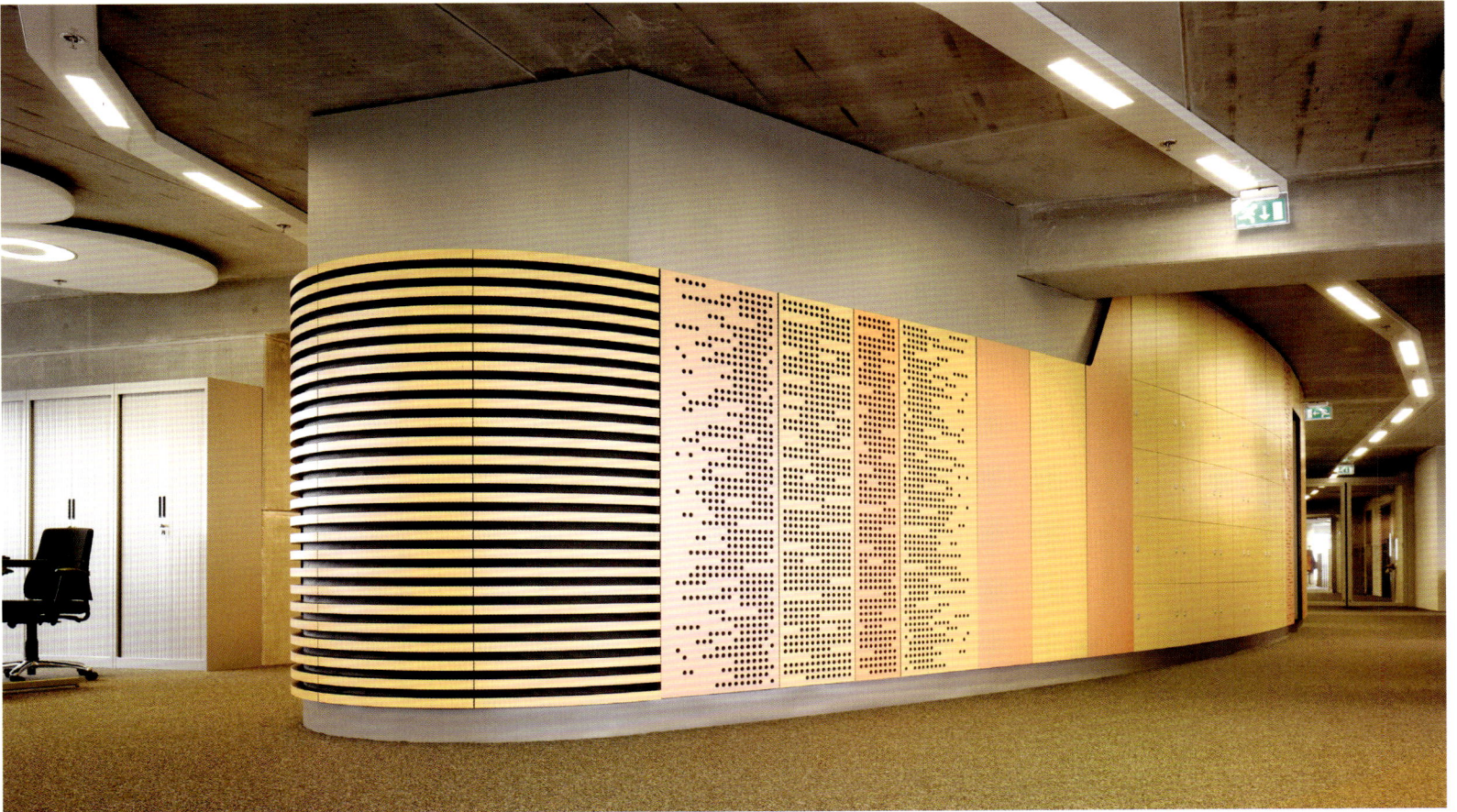

设计师没有简单地设计有死角的直线型走廊，而是为每条走廊设计了一条路线可以观赏大楼的景色。你可以在大楼里悠闲地散步，从透明玻璃向外领略周围的景色。设计中包含了许多的创新点，包括材料的减少、低能耗和许多包含可持续设计点的工作环境。

The layout of the building does not create simple linear corridors leading to dead ends, but instead each corridor has a route which introduces a kind of landscape into the building. You can take endless walks through the building, where there is a great deal of transparency, also towards the surrounding landscape.

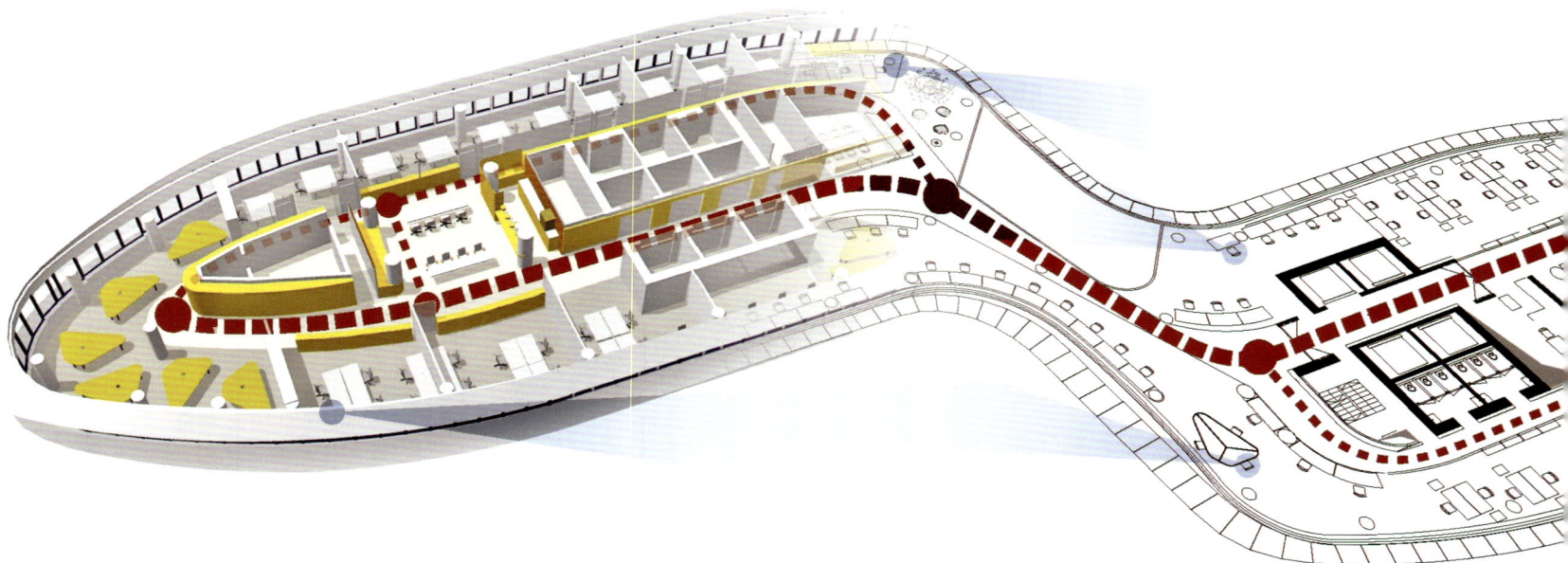

e)

n wide)

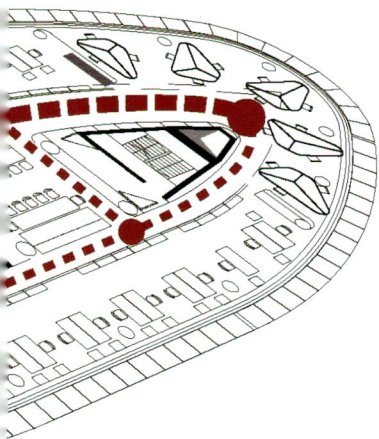

LANTANA CAMPUS

Lantana 创意园

设计公司：Ehrlich 建筑事务所、Wolcott 设计公司
项目地点：美国加州
摄影师：RMA Photography 公司、Marshal Safron

项目介绍　PROJECT DESCRIPTION

位于美国加州圣莫尼卡奥林匹克大道旁的 Lantana 创意园是一个知名的商业地产，其中的租客都是从事艺术创意产业的公司，占地达到 48 000 平方米。

Lantana

该项目的建筑使用了环保而经济的材料——自然色的喷砂水泥，使建筑能适用于住房、轻工业和商业设施的多行业运用。设计结合了宽阔的环保景观系统以及和谐的植被色调，使创意园建筑更具统一感。

Cost effective, low-impact materials such as a natural finish sand-blasted concrete combined with scaled-back forms, allow this development to assimilate within the varied context of residential, light industrial and commercial facilities. The campus design integrates an extensive, sustainable landscaping system and defined palette of plant materials with the building architecture to create a unified look for the overall development.

这个集中了高科技和影视后期制作产业的创意园分为一东一南两栋大楼，中间由一条人行道连接。

This high-tech, post-production campus includes two new structures, Lantana East and Lantana South, being connected via landscaped pedestrian zones.

Beachbody，作为一个生活、健身与营养方面的顾问公司和多功能中心，坐落在创意园的南楼，有两层，面积 3995 平方米，其中包括前台、一个互动休息室以及一个集研讨、会议和其他功能为一体的综合区。会议室使用了玻璃墙，创造了一个开放型的空间。

美国国立录音艺术科学学院（NARAS）坐落在东楼。两层楼共 6131 平方米，大厅中有一块显示器向来访者展示学校的光荣历史。该学院设有众多行政管理办公室，随处可见带有设计美感的图案，宣扬录音产业的魅力。

Beachbody, a lifestyle, fitness, and nutritional supplement company, occupies Lantana South and consists of a two-story, 3995-square-meter space with a multifunctional core that incorporates reception, an interactive lounge, integrating space for presentation, conferencing, and company functions. Use of Moveo glass walls on the conference rooms maximizes the capability for open space.

NARAS' offices include a two-story, 6131-square-meter space in the East. The office features a two-story lobby with a water wall element and an integrated display chronicling the history of recording. The facility, which accommodates the executive and administrative offices, has an aesthetic infused with iconic images that pay tribute to the recording industry.

两栋大楼的共同特点是阳台和可遮蔽的窗户既能让室内环境同室外紧密联系，又不影响挑剔的租户通过可调节太阳得热的窗户看向室外的视野。

Both buildings share a vocabulary of balconies and shaded windows that provide the interiors with a visceral connection to the outside while providing the high-profile tenants with unobstructed views through solar-gain controlled windows.

LARCHMONT OFFICE

Larchmont 公司办公室

设计公司：Rios Clementi Hale 设计工作室
项目地点：美国洛杉矶
项目面积：1581 ㎡
摄影师：Tom Bonner

项目介绍 PROJECT DESCRIPTION

Rios Clementi Hale 设计工作室是一个位于美国洛杉矶的多元化专业设计公司，从事包括建筑、室内设计、景观规划、平面、产品设计的工作。该公司的设计作品以睿智、实用、美观著称，2007 年曾获得美国建筑师学会加州分会设计公司奖，2009 年获美国国家设计奖提名。

Larchmont 公司办公室一层平面图   RCHS_Larchmont plan_Fl 1

Rios Clementi Hale 设计工作室的新办公室是由一家迷你商城改建的，外墙上镶有落地窗，门廊由玻璃和铝板组成，二楼工作室与屋顶平行，屋顶的周围开着天窗，提升了整个建筑物的线条，更好地融入城市街景。镜面玻璃和铝金属在二楼的镂空装饰板中若隐若现，花纹取材于传说中有着人身鱼尾的海神的脸，同这条大街上的一所著名的艺术学院的标志相映衬。

Renovation of the former mini-mall includes replacing the exterior walls with a near floor-to-ceiling window wall system, creating porches of glass and aluminum panels—inspired by the Gamble House sleeping porches—around the second floor studio space, and leveling off the roof surrounding the existing skylights, to upgrade and streamline the space to fit the building into the sophisticated city streetscape.The panels on the second floor alternate between mirrored glass and cut aluminum that display an abstracted design of the face of Triton, which pays homage to the California Swim Arts School that was once located at 639 Larchmont.

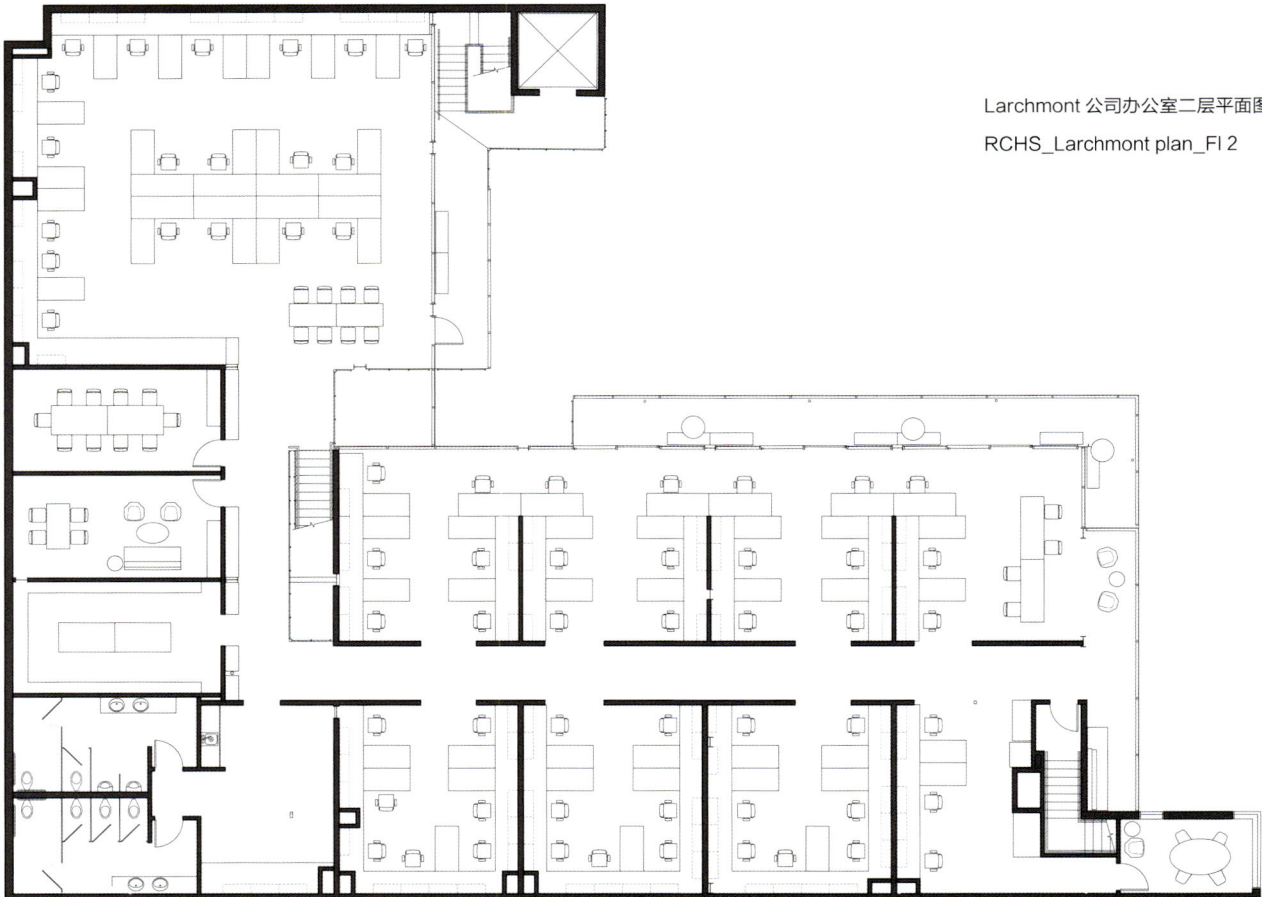

Larchmont 公司办公室二层平面图
RCHS_Larchmont plan_Fl 2

除分布在每一层的工作室外，底层包含了 930 平方米的 notNeutral 零售店，里面展示了公司设计的产品；同时它还经营批发业务，有一个沙龙、一个咖啡厅和商店。设计师希望将销售与生活情趣设计和谐地融合在一起，形成对公众开放的友好空间。

In addition to the 12000 square feet of studio space located on each level, the ground floor comprises the 1000-square-foot notNeutral retail flagship store, including the product company's design and wholesale operations. Additional retail space is occupied by a salon, as well as a yet-to-be-named café and retail shops. The architects desire a harmonious mixture of retail businesses that promotes great lifestyle and design that is inviting to the public.

办公区淡化了等级划分，整个设计依照现代设计理念，创造出一种流畅的布局、自然的配置。六到十九个工作台可移动，便于员工位置的变换。管道和天花板结构裸露在外，同箔片交织在一起制造出一种工业感，将"概念工厂"的创意引入办公室。

The exposed ducts and structural systems with the use of foil at the ceiling create an industrial feeling to the space, recalling the foil-lined studio of Andy Warhol, thus bringing the firm's idea of office as "idea factory" to life.

巴罗达投资公司

设计公司：Rios Clementi Hale 设计公司
项目地点：美国加州
项目面积：388 ㎡
摄影师：Tom Bonner

项目介绍 PROJECT DESCRIPTION

巴罗达投资公司由慈善企业家 David C. Bohnett 于 1998 年在美国洛杉矶建立。这位慈善家曾建立过目前已被雅虎公司收购的 GeoCities 公司。巴罗达主要的投资方向是网络、软件服务、数字媒体。

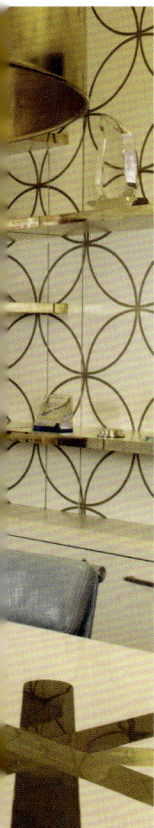

一楼会议室的小窗户用四叶花瓣的铁条装饰，这种图案贯穿整个设计主题。金色四叶花瓣装饰了会议室里白色的墙纸，天花板上的圆形石灰浮雕上挂着三盏黄铜装饰灯，与光滑的白色长桌上的星号图案完美搭配，并配有埃姆斯鳄鱼皮椅。

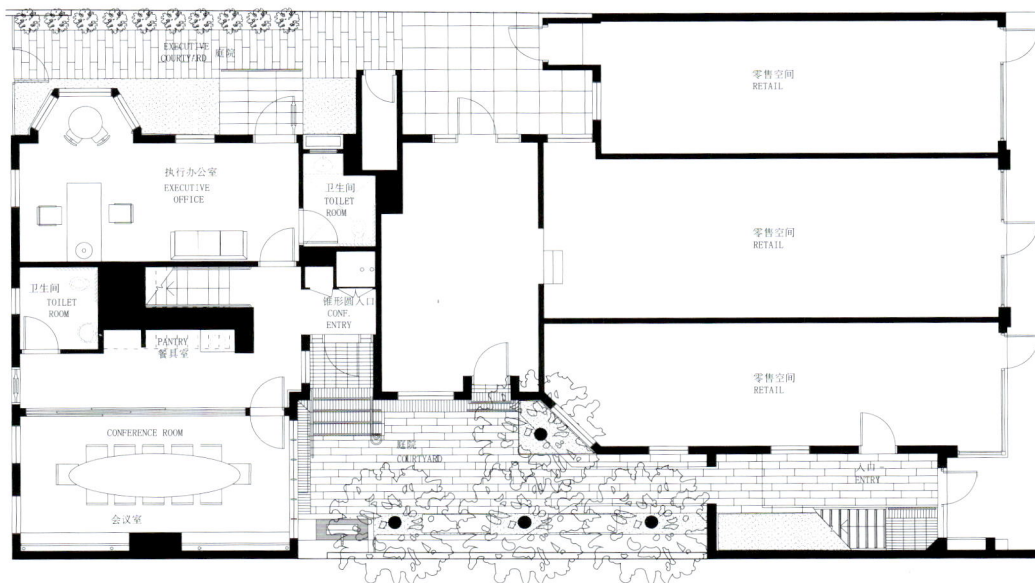

设计师成功地将复古和现代风格混合在这个两层办公室的设计中。现代经典的家具搭配风格迥异的软装、精美的天花板饰品、门上夸张的猫眼、高光漆面、变换大小的重复性花纹，在充足的光照下，美轮美奂，升华出超越时间的优雅。

A custom-designed steel-and-glass staircase provides access to the second floor, affording uninterrupted views of the fountain below. At the top of the staircase, through a wooden door adorned with an exaggerated escutcheon, is the reception area, featuring a two-seat Eames Sofa in striped upholstery and an aqua leather Barcelona Chair. The ceiling is removed and seismically restructured to expose the existing hip roof. Interior walls are clad in either white back-painted glass, limed oak panels, or modern versions of tongue-and-groove, which function as textural foils for the existing masonry shell.

钢制和玻璃的定制楼梯，直通二楼，从上面可以看到底层喷泉全景。来到二楼，推开木门，映入眼帘的是接待区，有两个条纹埃姆斯沙发，一把浅绿色的皮质巴塞罗那椅。以前的天花板被拆掉，改成了现在的坡形抗震结构。原有内墙的石板上镶上了黑白涂色玻璃、石灰橡木板以及现代的接合沟槽，作为现在砌体壳真正的保护膜。

Rios Clementi Hale Studios adapted a surprising, yet delightful combination of retro and contemporary styles for the renovation of the two-storey Baroda Ventures office. The architects applied several themes throughout the design—classic modern furnishings with unexpected fabrics, elaborated ceiling medallions and doors escutcheons, glossy surfaces, and repeated patterning at various scales—while incorporating plentiful daylight. Timeless elegance was achieved by pairing traditional essentials with the latest design to create a place of sophisticated opulence.

AUTOBAN

Autoban 设计公司

设 计 师：Alexander Valluri
设计公司：Autoban 设计公司
项目地点：土耳其伊斯坦布尔
摄影师：Ali Bekman

项目介绍 PROJECT DESCRIPTION

Autoban 是位于土耳其伊斯坦布尔的一家新设计公司。在 100% Design 设计展（伦敦）上，这家公司的设计师获得 Blueprint Award 的最佳新人奖。他们的设计既有一种如同伊斯坦布尔给人的古老神秘感觉，也有注重功能的现代感，并把两者很好地融合起来。

"

The Autoban team of 30 is now settled in a brand new office in the heart of Istanbul. Designed by the Levantine architect Alexander Valluri, this late 19th century building was originally a Union France Bank. Original features such as ornate wooden framed windows and marbled flooring were restored by Autoban, and the deserted space is now neatly divided into workspaces, whilst through the use of transparent and white materials, the grandeur of the space is preserved. The result is a fresh environment, truly a pleasure to work in.

AUTOBAN 办公区平面图

有 30 位成员的 Autoban 设计公司坐落在伊斯坦布尔的中心。这座古老的 19 世纪建筑，原本是法国联盟银行的所在地。Autoban 恢复其华丽的原始特征，像木制框架窗口和大理石地板。这个曾经被遗弃的空间如今被划分成整齐的工作区。透明的白色材料保留了空间的美感。其结果是让员工在这样清新的环境中享受工作的乐趣。

Wallpaper* Design Awards 2004
Best Young Designer

BLUEPRINT 100% Design 2005
Best Newcomer

Best New Restaurant
Müzedechanga

ELLE DECO International Design Awards 2007
Turkey Furniture Category Winner with Table Table

ELLE DECO International Design Awards 2008
Turkey Lighting Category Winner with Dying Spider
Turkey Bedroom Category Winner with Bergère Bed

DESIGN TURKEY 2008 Design Awards
Superior Design 2008 - Bergère
Good Design 2008 - Woody
Good Design 2008 - Arm...

东方ＩＣ创意办公室

设计师：Thomas Dariel & Benoit Arfeuillere
设计公司：Lime 388 by Dariel and Arfeuillere
项目地点：上海
项目面积：700 ㎡

项目介绍 PROJECT DESCRIPTION

东方 IC 是一家专业的多元化视觉资源提供商和图片技术服务商，公司成立于 2000 年，是东方网入股企业。公司拥有新闻图片中心、创意图片中心、东方 IC 包年、英语图片网站这四个图片平台，以及东方 IC 图片管理系统、东方 IC 视觉传播等增值服务，东方 IC 的产品和服务覆盖了从专业到大众，从低端到高端，从新闻到创意，从国内到国外的整个图片行业领域。

七米高玻璃屋顶的摄影棚位于中心区域，象征着公司的核心业务。它使整个空间有机统一，组织引领着办公室里其他工作区域。拍摄和图片编辑是一项需要安静环境的工作，因此我们采用了天然材料并混合了水泥和木头，运用明亮的白色并结合了风水创造出了"禅意"花园。

The shooting studio with its seven meters ceiling and its glass roof is central located. Standing for the core business of the company, it brings light to the entire space and organizes the rest of the office that gravitates around it. Shooting and editing were tasks requiring a peaceful environment and thus are designed accordingly by preferring natural materials with a mix of concrete and wood, bright white color and adding a fengshui touch while creating a "zen" garden.

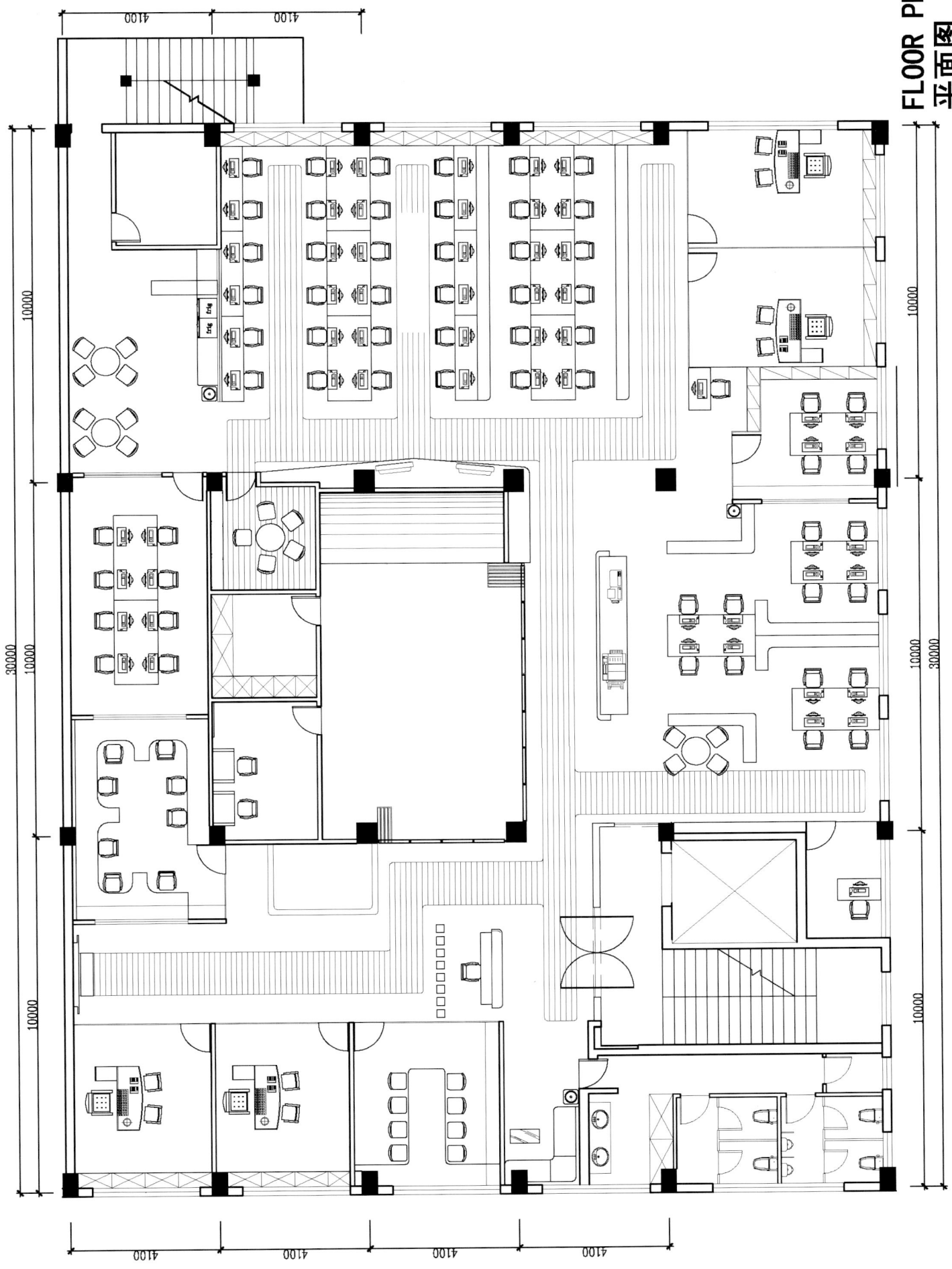

开放式工作区配备了大型白色亮漆办公桌，营造了友好的工作氛围，使员工交流起来更轻松。配色方案也仍然大面积选用纯净的白色。而会议室的玻璃门选用了红色以平衡并打破白色的单调。同样，在图书馆区，几组暗色调几何型书架具有强烈的画面感和视觉元素。

The open working space is furnished with large white lacquer desks and answers the requirement of friendly atmosphere and easy communication between the staff. The color scheme again is mainly pure white. Yet a bright red color applied on the meeting rooms, doors engenders contrast and balance while breaking the all-white rhythm. Likewise in the libraries, several groups of dark geometric open bookcases constitute strong graphic and visual elements.

JWT NYC

智威汤逊（JWT）纽约总部办公室

设计公司：Clive Wilkinson 建筑公司
项目地点：美国纽约
项目面积：23250 ㎡
摄影师：Eric Laignel

项目介绍　PROJECT DESCRIPTION

智威汤逊（JWT）创始于 1864 年，是全球第一家广告公司，也是全球第一家开展国际化作业的广告公司。自成立以来，智威汤逊一直以"不断自我创新，也不断创造广告事业"的理念著称于世。它以品牌全行销规划，结合广告、直效行销、促销、赞助及公关活动，致力于协助客户达成短期业绩成长，并创造长期的品牌价值。 时至今日，智威汤逊风采依旧，昂首跻身于世界四大顶尖广告公司之列，目前隶属于全球最大的传播集团 WPP。

设计师为 JWT 公司设计出一个代表新兴的经营模式和品牌形象的富有创造力、灵活、开放的空间。设计的主题是以"讲故事"的"树木"为中心，将空间内每个分支机构串联为一个整体，形成形式统一且可以互相交流的空间。

The overall design goal is a creative, flexible and open architecture to reflect its revitalized business model and brand identity. We use the tree as a metaphor for storytelling and extend it as an organizing form and connective tissue between the individual branches of the agency.

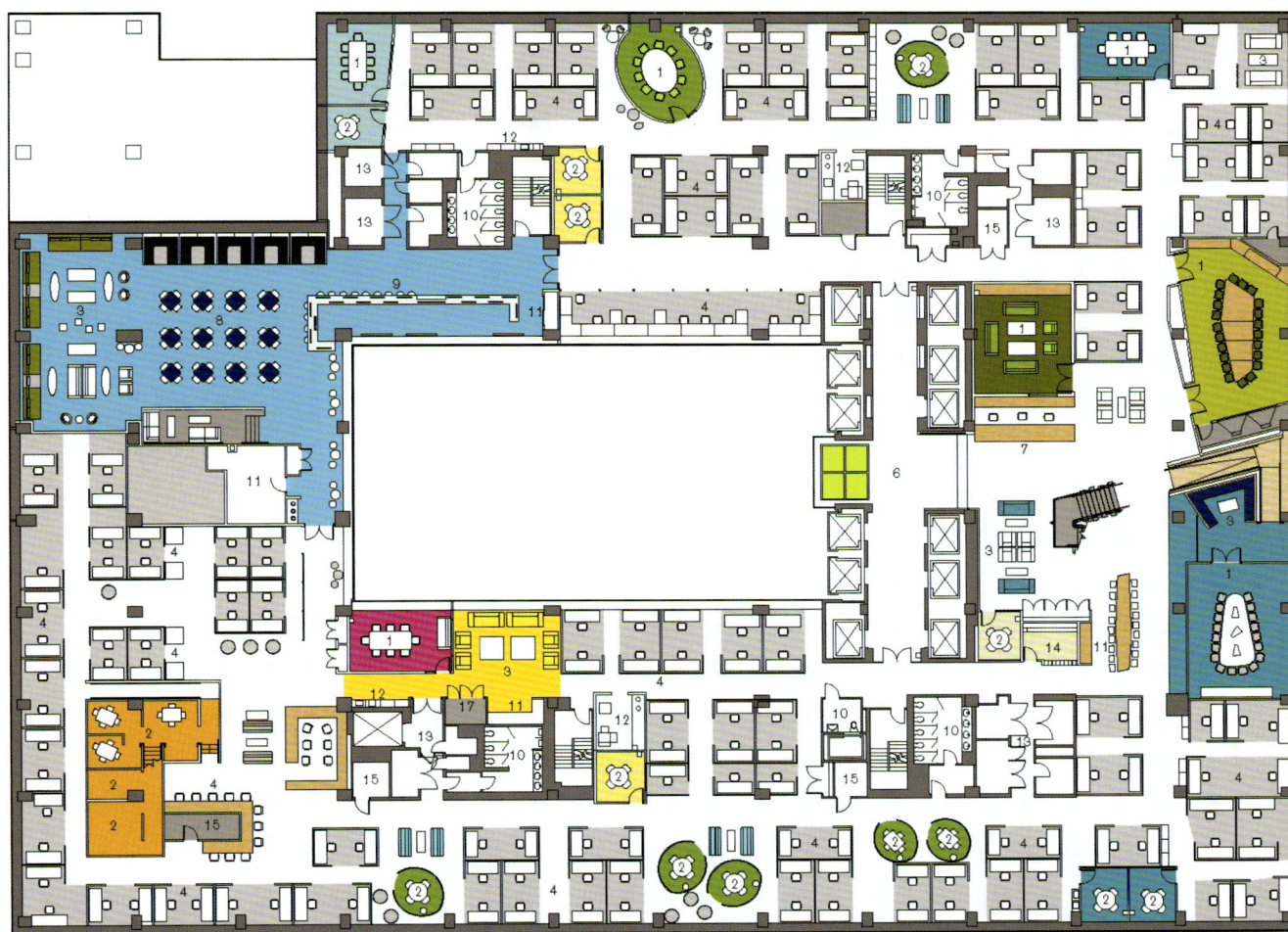

01 会议室
02 小型会议室
03 休息室
04 开放式工作站
05 办公室
06 电梯间
07 接待室
08 集会厅 / 咖啡室
09 酒吧
10 休息室
11 餐具室
12 复印室
13 服务系统
14 收发室
15 储藏室
16 视频 / 音频出品室
17 信息技术室

01 Conference Room
02 Small Meeting Room
03 Lounge
04 Open work Stations
05 Office
06 Elevator Lobby
07 Reception Area
08 Town Hall / Cafe
09 Bar
10 Restrooms
11 Pantry
12 Copy / Print
13 Services / Systems
14 Mail
15 Storage
16 A/V Production
17 IT

JWT 公司总部，纽约   二层平面图   JWT Headquarters, New York
Level 02

0'   10'   20'   40'

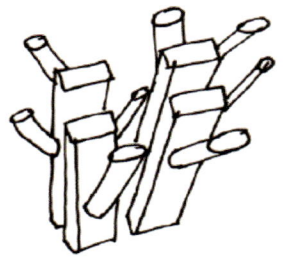

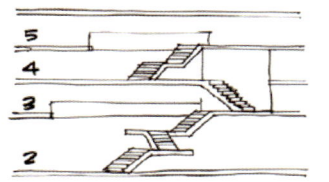

> 16 个不同的"树干"上有数控雕刻的字母，上面有"很久以前"的字样，像是一本名著里的开篇。单词部分镂空，切下来的字母垂下来，像一片片树叶。每个句子都是不完整的，读者可以自由想象。"树枝"构成了一个个卵圆形的会议室，墙面有一定的角度，像一条条拉升穿过楼层的"树枝"。

The sixteen different tents extend the metaphor further by each character incised using CNC machines with the first sentence of a famous novel. The words are cut into the fabric and the cut letters hang down appearing like leaves. Each sentence is incomplete in meaning, so readers are free to extend its meaning. The branches are ovoid-shaped meeting rooms and the cones are angled, like branches stretching through floors.

THE FOURTH LEVEL

小型会议室
▼ SMALL CONFERENCE ROOM

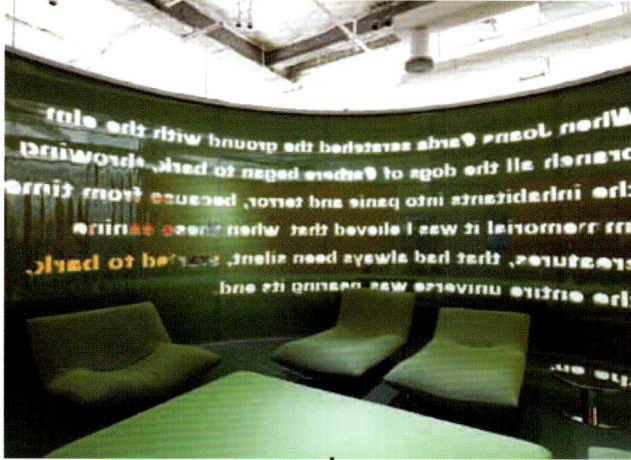

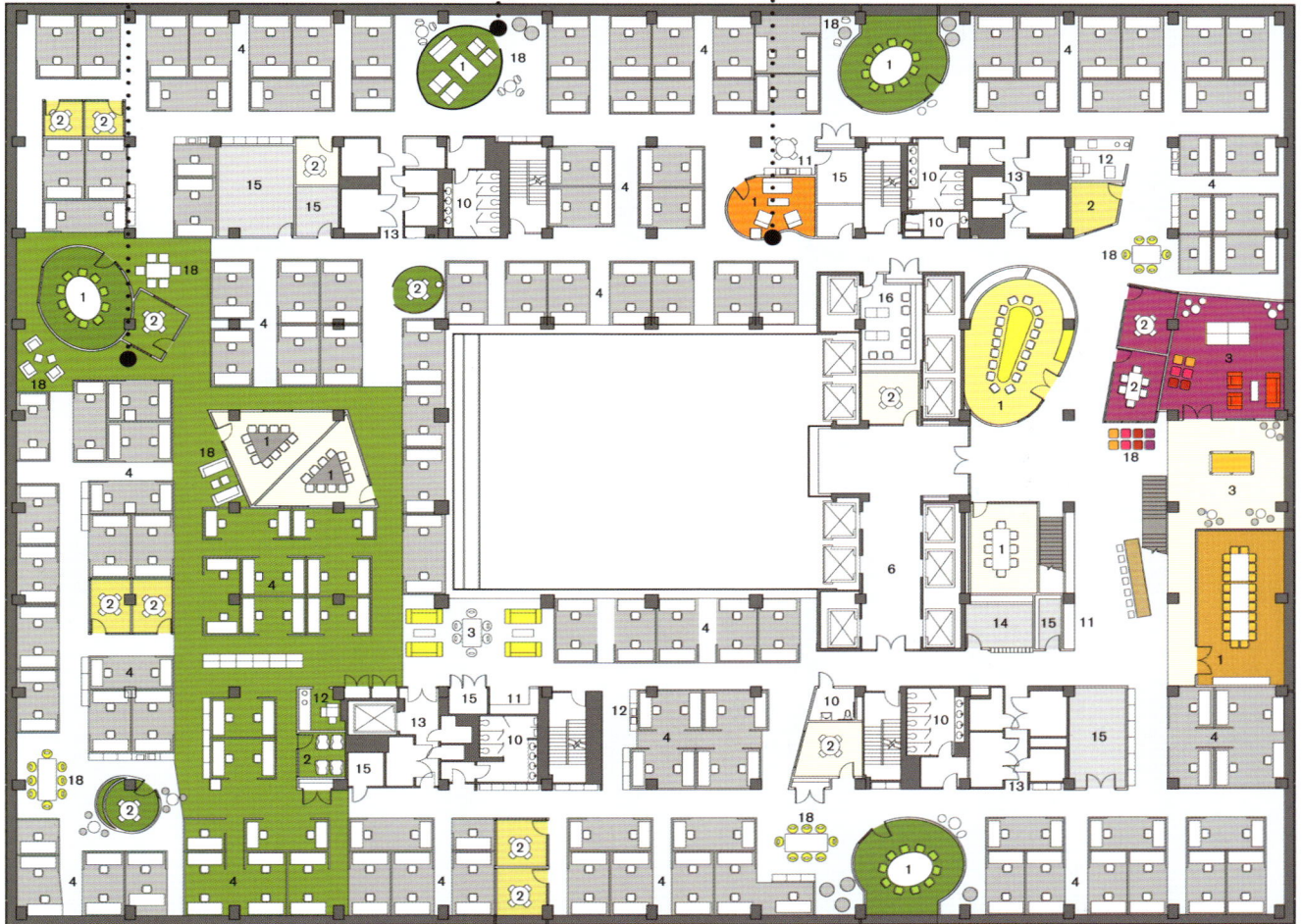

会议室
▼ CONFERENCE ROOM

会议室
▼ CONFERENCE ROOM

01 会议室
02 小型会议室
03 休息室
04 开放式工作站
05 办公室
06 电梯间
07 接待区
08 集会厅 / 咖啡室
09 酒吧
10 休息室
11 餐具室
12 复印室
13 服务系统
14 收发室
15 储藏室
16 视频 / 音频出品室
17 信息技术室
18 休息室

01 Conference Room	10 Restrooms
02 Small Meeting Room	11 Pantry
03 Lounge	12 Copy / Print
04 Open work Stations	13 Services / Systems
05 Office	14 Mail
06 Elevator Lobby	15 Storage
07 Reception Area	16 A/ V Production
08 Town Hall / Cafe	17 IT
09 Bar	18 Break room

0'   10'   20'   40'

入口中庭约 9.7 米高，一座混凝土做的楼梯贯穿其中。二层的精彩设计得益于 5.5 米高的天花板。整个空间都没有吊顶，小型奥德堡照明系统为这里提供了足够的光线，减少了室外彩色窗户遮挡阳光的负效应。

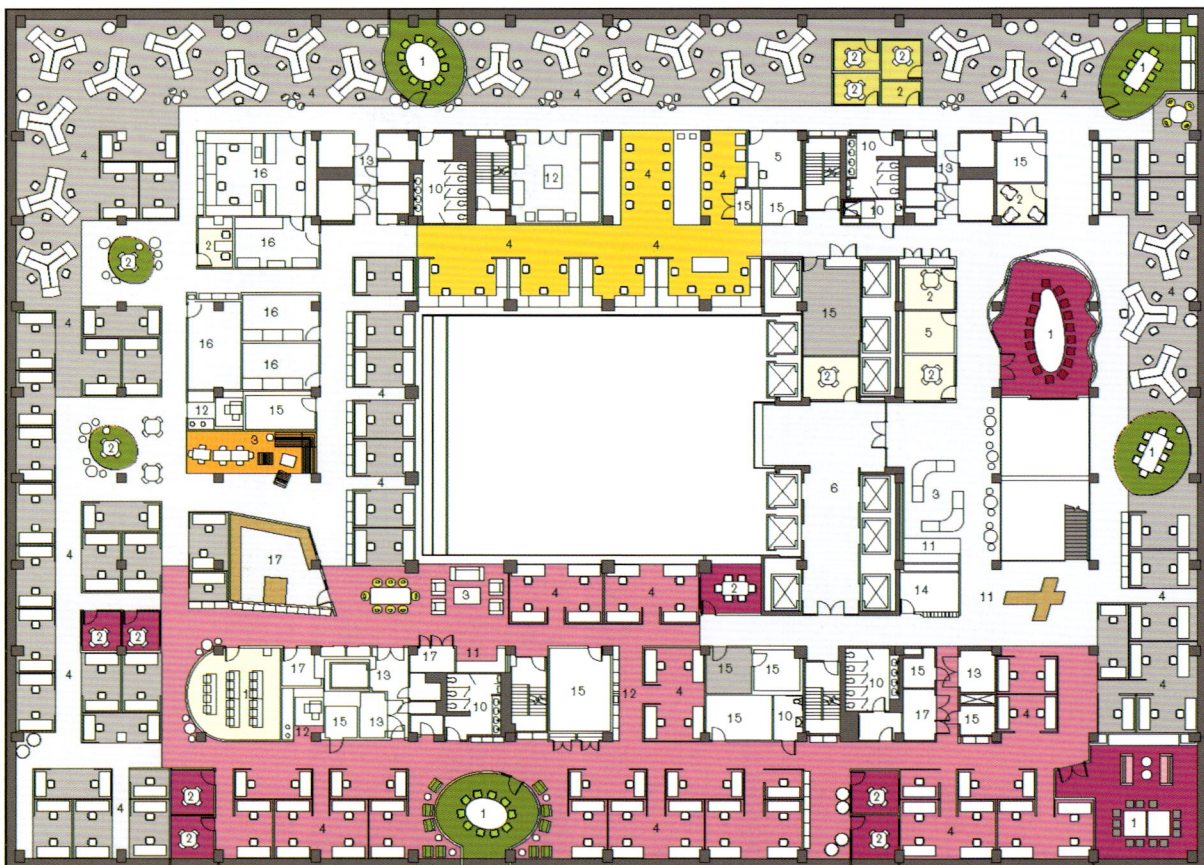

01	会议室
02	小型会议室
03	休息室
04	开放式工作站
05	办公室
06	电梯间
07	接待室
08	集会厅 / 咖啡室
09	酒吧
10	休息室
11	餐具室
12	复印室
13	服务系统
14	收发室
15	储藏室
16	视频 / 音频出品室
17	信息技术室
18	休息室

01	Conference Room
02	Small Meeting Room
03	Lounge
04	Open work Stations
05	Office
06	Elevator Lobby
07	Reception Area
08	Town Hall / Cafe
09	Bar
10	Restrooms
11	Pantry
12	Copy / Print
13	Services / Systems
14	Mail
15	Storage
16	A / V Production
17	IT
18	Break room

JWT 公司总部，纽约　　五层平面图　JWT Headquarters, New York
Level 05

0'　10'　20'　　40'

主入口的楼梯通往每层楼。随着楼梯的延伸，我们进入到具有不同特色的邻近区域，特制的路标利于激发寻路的兴趣。最具创造性的区域包括中间楼层的"树房子"、会议厅、咖啡厅以及布满灯饰的酒吧。

The staircase connects all floors over the main entrance hall. As circulation of the stairs flows into distinctive neighbors representing various departments within the agency, special landmarks help create visual interest while facilitating way-finding interest.

GOOGLE

Google 墨西哥办公室

设计公司：Space 设计公司
项目地点：墨西哥城
项目面积：850 ㎡
摄影师：Willem Schalkwijk

项目介绍 PROJECT DESCRIPTION

创始人 Larry Page 和 Sergey Brin 在斯坦福大学的学生宿舍内共同开发了全新的在线搜索引擎，然后迅速传播给全球的信息搜索者。Google 目前被公认为是全球规模最大的搜索引擎，它提供了简单易用的免费服务，用户可以在瞬间得到相关的搜索结果。作为世界领先的网络搜索平台，Google 在墨西哥城建立了一个充分反映公司特色的办公点。"网页浏览体验"就是设计理念出发点。

"

Once inside, color is obvious; different layers make you discover new things as you walk by. Designed as a whole, what happens on the ceiling has a direct repercussion on the floor and walls.

Open work space talks about productivity and concentration in the middle of a really busy environment. A central round meeting room that represents the heart of the design, and around which other meeting rooms and private offices rotate. Then you find a yellow box, a zen like huddle that serves as the transition into the informal work space.

> When you enter the reception you find it a really white space, with a bright-colored back wall and a white floor; wall and ceiling as a never ending wrapping represents the first encounter to the web. Here you can look for what you want, and it is only when you go through the black glass doors that you can really find the different ambiances/options that you have to choose from.

该办公室采用低挥发性材料、环保地毯和木料，并设置了垃圾回收箱。整个空间设计简单，造价低廉，又营造了活泼有序的氛围。

空间布局分为三部分：带工作台、会议室和私人办公室的正式工作区，简单设计的非正式聚集区、休闲区，一个巨大的可观赏外景的露台。前台的白色墙纸、地板、天花板象征了初次进入网络世界的心情，与黑色玻璃门后五彩斑斓的世界形成对比。进门后的天花板、墙面、地板成为一个整体。办公室中间的圆形会议厅被设计成其他会议室和私人办公室的轴心。黄色的禅聚地是向非正式工作区域过渡的空间。

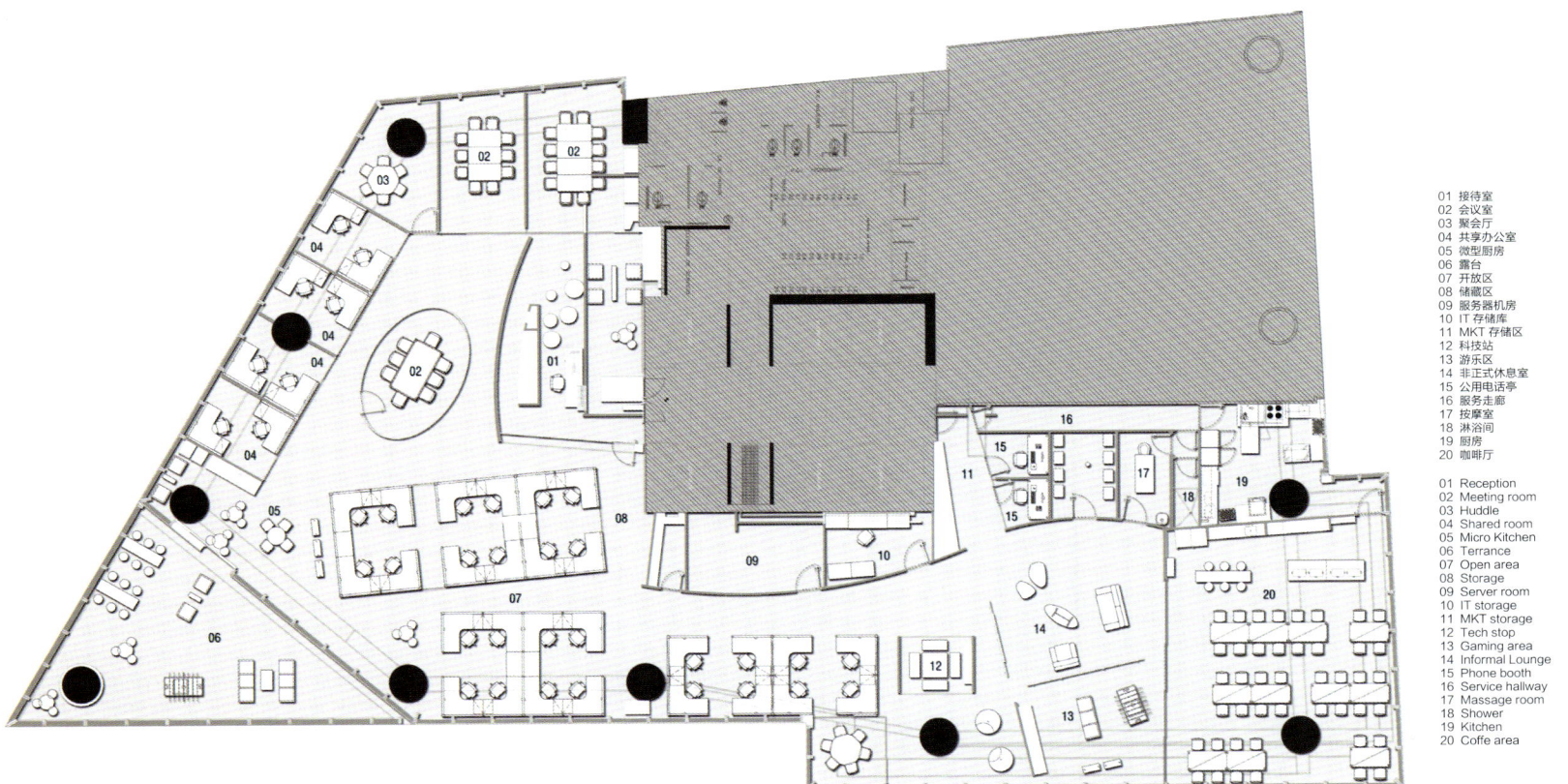

01 接待室
02 会议室
03 聚会厅
04 共享办公室
05 微型厨房
06 露台
07 开放区
08 储藏区
09 服务器机房
10 IT 存储库
11 MKT 存储区
12 科技站
13 游乐区
14 非正式休息室
15 公用电话亭
16 服务走廊
17 按摩室
18 淋浴间
19 厨房
20 咖啡厅

01 Reception
02 Meeting room
03 Huddle
04 Shared room
05 Micro Kitchen
06 Terrance
07 Open area
08 Storage
09 Server room
10 IT storage
11 MKT storage
12 Tech stop
13 Gaming area
14 Informal Lounge
15 Phone booth
16 Service hallway
17 Massage room
18 Shower
19 Kitchen
20 Coffe area

LENOVO

联想集团墨西哥办公室

设计师：Juan Carlos Baumgartner、Jimena Fernández Navarra
设计公司：Space 设计公司
项目地点：墨西哥
摄影师：Santiago Barreiro

项目介绍　PROJECT DESCRIPTION

联想集团是一家计算机技术跨国公司，以开发制造桌面和手提电脑、控制台、服务器、存储驱动盘、IT管理软件和相关业务为主。而今，联想的主要运营点在中国北京和美国莫里斯维尔，并在世界各地都有分公司。

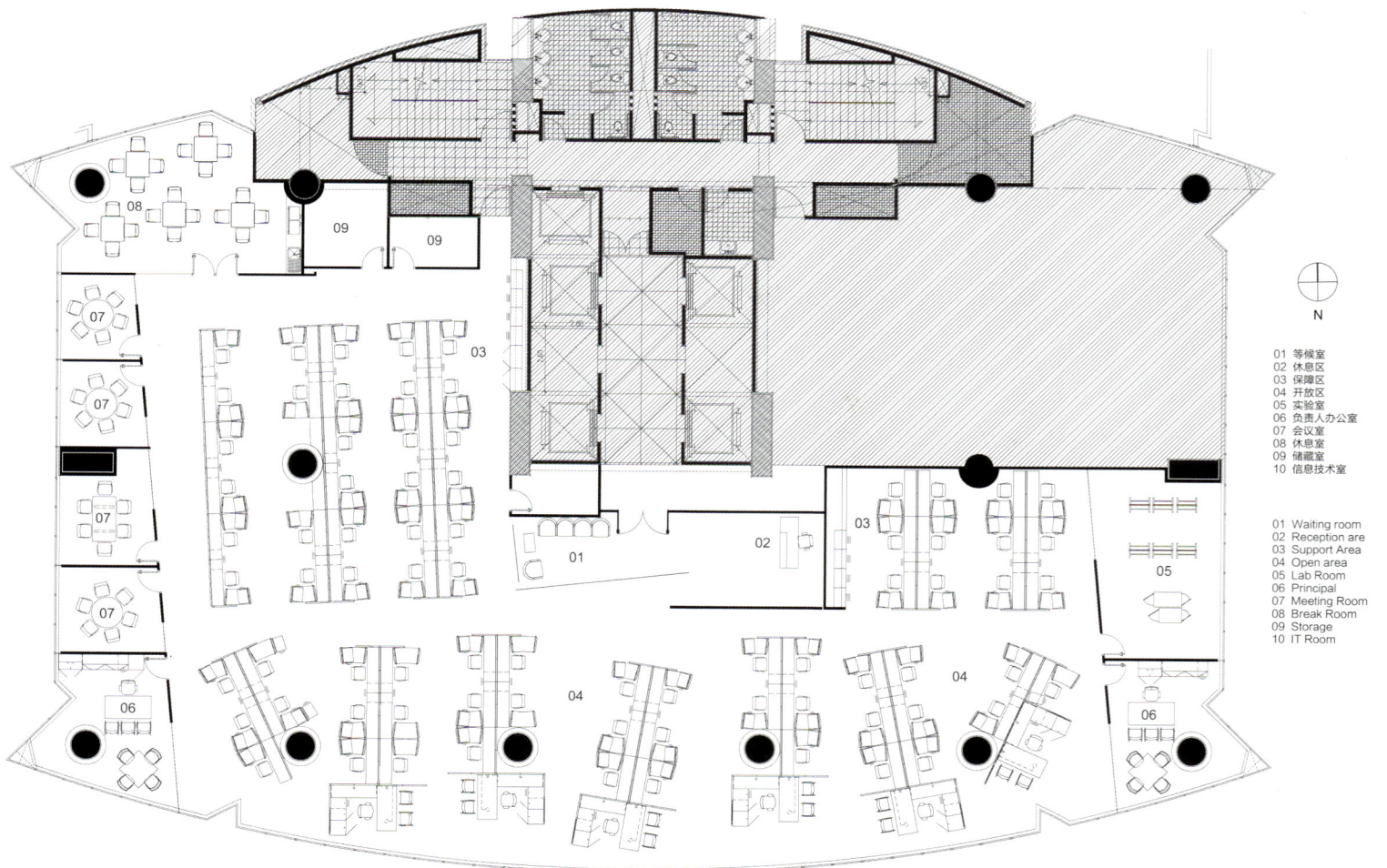

01 等候室
02 休息区
03 保障区
04 开放区
05 实验室
06 负责人办公室
07 会议室
08 休息室
09 储藏室
10 信息技术室

01 Waiting room
02 Reception are
03 Support Area
04 Open area
05 Lab Room
06 Principal
07 Meeting Room
08 Break Room
09 Storage
10 IT Room

联想的员工大多都是时尚而朝气蓬勃的年轻白领。整个设计理念是设计一条贯穿整个办公室的过道，石灰板和地毯布置在工作区的边围，过道的尽头有一间实验室，来访的客户在进入公司时，就能看到品牌标识。工作区主要分布在长方形空间的短边上，透明挡板增加了空间左边的光照，同时也腾出更多空间让长边进光。为了让空间设计更美观，所有的服务设备都隐藏在大楼的后部和盲区；为了让空间更明亮，照明系统采用工厂照明使用的最小尺寸的线性光源。有些地方的光线能达到 2.6 米长。根据不同用途，不同的区域的照明设备也会被安置在不同的高度。照明系统达到欧洲标准，符合人体舒适度。

Design for Lenovo is based on an organized chaos. It is a design for people who work here, who are young, and for the company, which is urban, trendy and fun. All the concept is based on a straight path that leads us through an area of chaos, where the drywalls and carpets swivel away from the workstations, and where the end of the path is the lab room, a place where the customers will be shown while getting acquainted with the rest of the offices, so that they get a general understanding of the brand's concept. To get the best out of the scheme, all services are sent to the rear and blind area of the building. In order to illuminate this space, a special illumination is designed for the project, creating continuous lines of lights of minimum size with 100% industrial look. At places where lines are suspended at 2.60, the drywall is intercepted where the line of lights is embedded. Levels of lighting are separated in different areas depending on to their uses, according to European standards and the perception of the human being.

WE ARE THE NEW.
WE ARE THE FUTURE.
WE ARE LENOVO.

INNOVATION FOR PLAYING HARD

办公室的天花板很高，于是我们将门窗设计到 2.6 米高，顶上用金属架制造了一个灯箱，在强调立面的同时，给空间带来更多工业氛围。虽然空间的格局很开放，但设计师利用创新与有活力的设计元素巧妙地分出了工作等级。

Having a very generous height, we tried to give it a more human scale in the front of the built spaces by placing doors and windows of 2.60 meter in height, and making a light box with metallic grid from the upward, which is thought to emphasize the façades, and in turn provide the space with an industrial atmosphere.

POWER OFFICE

荷兰阿姆斯特丹股票交易办公室

设计公司：i29 l interior architects & eckhardt&leeuwenstein architects
项目地点：荷兰
项目面积：240 ㎡
摄影师：i29 l interior architects

项目介绍 PROJECT DESCRIPTION

位于荷兰阿姆斯特丹 Herengracht 河畔的这家证券交易办公室是荷兰最大的证券交易公司之一，他们的投资范围囊括各个领域，但最主要是从事证券市场的交易。

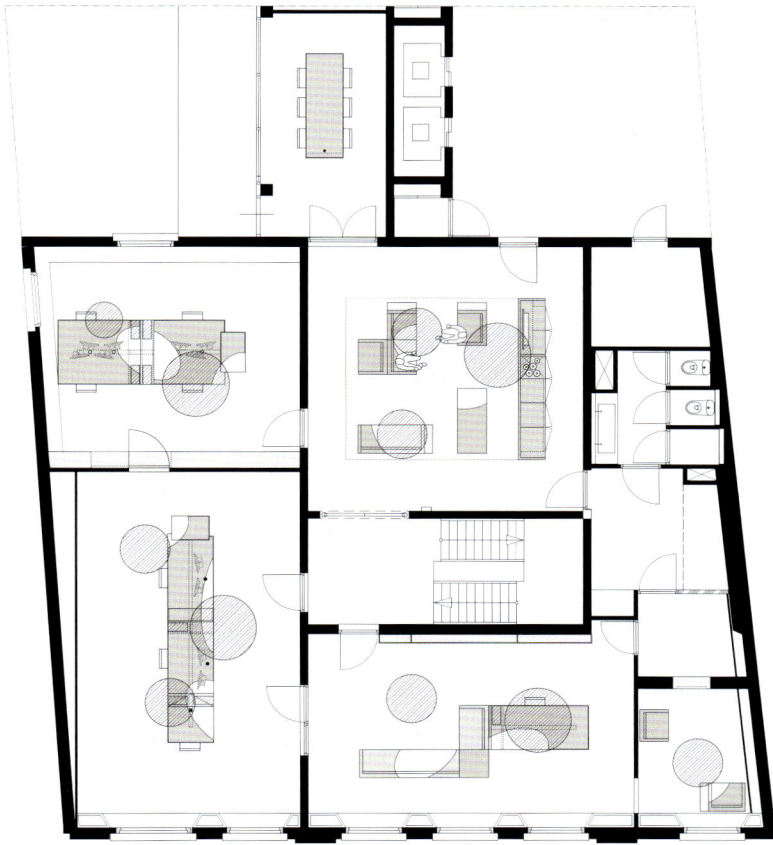

休息室的地板由白色大理石和颜色深浅不一的地毯拼接而成，吧台和铺着银色织物的长椅也使用了相同的设计手法。一整块平板做成的吧台和各类家具显现出的设计点显示这里可用于业务陈述或社会工作服务。

The lounge area has, in combination with the white marble flooring these same light/shadow patterns that cover the bar and benches in silver fabrics. This area can be used for presentations or social working, with an integrated flat screen in the bar and data connections in all pieces of furniture. The existing space is set in a 17th century historic building, at one of the most famous canals of Amsterdam called 'de gouden bocht'. All existing ornaments and details are painted white.

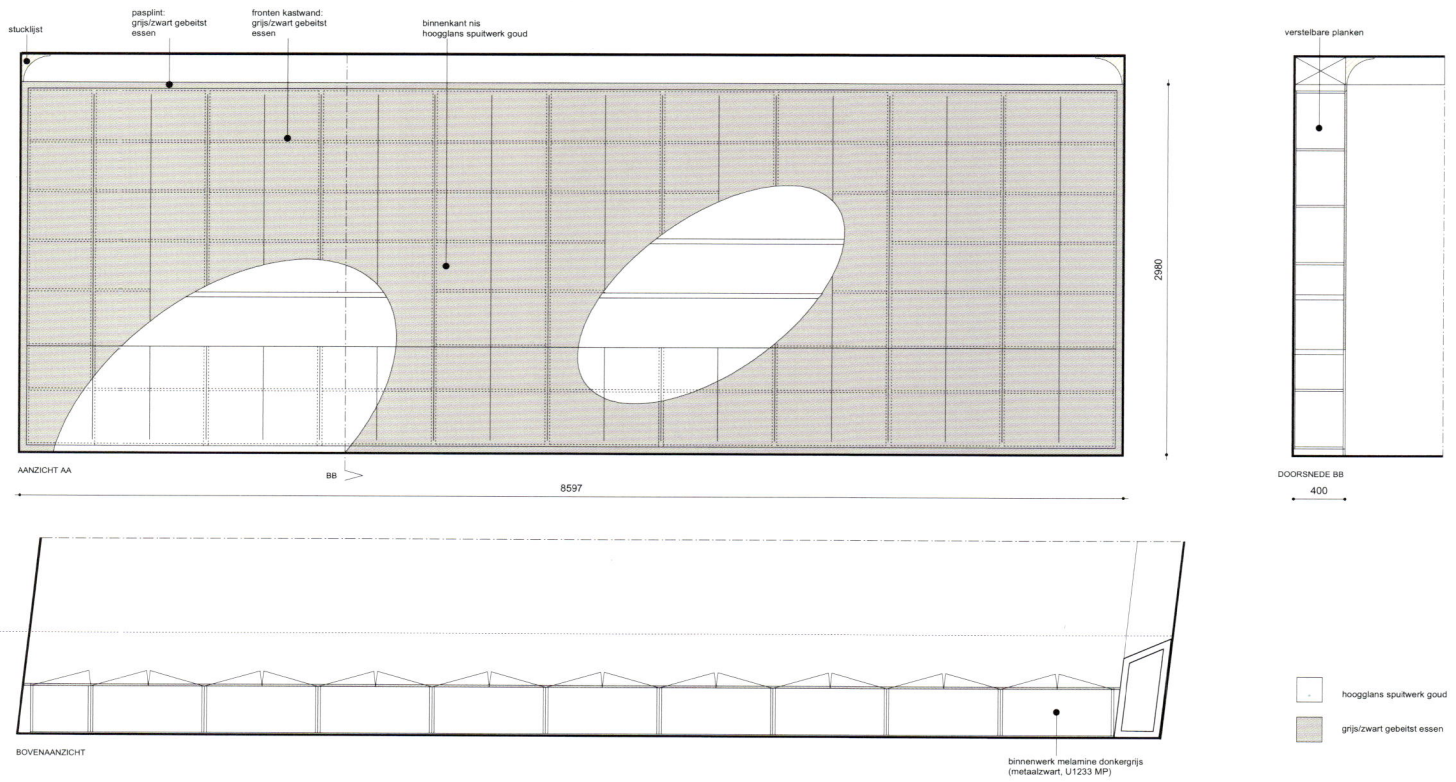

stucklijst

pasplint:
grijs/zwart gebeitst
essen

fronten kastwand:
grijs/zwart gebeitst
essen

binnenkant nis
hoogglans spuitwerk goud

verstelbare planken

2980

AANZICHT AA

BB

8597

DOORSNEDE BB

400

ST B

BOVENAANZICHT

binnenwerk melamine donkergrijs
(metaalzwart, U1233 MP)

hoogglans spuitwerk goud

grijs/zwart gebeitst essen

三个会议室和一个休息室展现总体设计理念，圆形的大灯罩，内表面是金色喷漆，为整个空间创造出具有光影效应的椭圆色标记。同时，金色的椭圆外形充满趣味，与有棱角的灰黑色白蜡木柜子和桌子形成鲜明的对比。地板上铺着深、浅灰色的地毯，勾勒出椭圆的形状，区分了不同的工作区。

All three boardrooms and a lounge are executed in an overall design concept. Large round lampshades, spray painted gold on the inside, seem to cast light and shadow of oval marks throughout the whole space. By this, a playful pattern of golden ovals contrasts with the angular cabinets and desks, which are executed in black stained ash wood. In the flooring the oval shaped forms continue by using light and dark grey carpet. Also, these ovals define separate working areas.

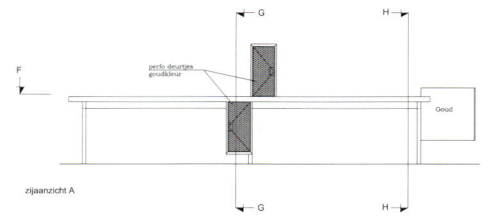

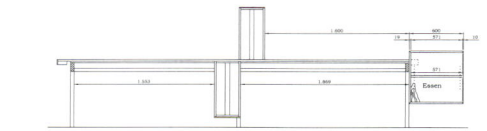

zijaanzicht A

doorsnede E - E

aanzicht D

aanzicht B

aanzicht C

aanzicht A

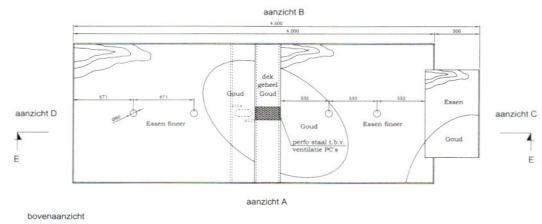

bovenaanzicht

"

该公司的关键词是金钱和力量。设计师很好地突出了这一点，将董事会每一位成员都安排在显著位置，金银色的椭圆形像一个个金币，打破了空间格局，暗示着这里是进行投资和股票交易的场所。

The keynotes for this company are money and power. The design concept expresses this by setting all members of the board literally in the spotlights. The golden and silver ovals shatter through the spaces like golden coins.

TIBA 物流公司西班牙办公室

设计公司：Space 设计公司
项目地点：西班牙
项目面积：1800 ㎡
摄影师：Courtesy of Organitec

项目介绍 PROJECT DESCRIPTION

TIBA 成立于 1975 年，总部设在西班牙的瓦伦西亚，是西班牙罗美欧集团 (Group Romeu) 全资子公司。罗美欧集团 (Group Romeu) 是西班牙一家非常有实力和影响力的全球化物流公司，至今已有 100 多年的历史。作为西班牙物流业的领军角色，罗美欧集团旗下拥有 30 多个子公司一齐致力于物流行业各个环节。

设计的亮点之一是室内设施都露在外面，使天花板显得更高。不同形状和颜色有助于区分正式工作区和非正式工作区。玻璃墙让自然光线进入中心区域，减少对人工照明的依赖，这样既节能又能使员工沐浴阳光，享受风景。

电话亭为私人服务，方便他们处理个人事宜。接待客户或者召开会议的会议厅墙体用玻璃制成，天花板较低，隔音效果好。每个区域的椅子颜色不同，便于区分不同的用途。例如，运营区的椅子是橙色的，具有鼓励性；绿色的墙面在集中和力量间取得平衡；而会议室的颜色是蓝色的。讨论区多配以对比强烈的两种色调，如红色和白色，营造出与其他区域截然不同的氛围。

One of the most conspicuous points to look at was its leaving the installations exposed so that we could achieve better height and differentiate formal and informal work areas with different forms and colors. Phone booths were created for personal use in case they would require privacy and meeting rooms that possess glass walls and low ceilings with better acoustics for customer care or for meetings . Every area has chairs of different colors that distinguishes it for its attributes, for example, in the area of operations orange is chosen, a color that motivates; with green walls to achieve a balance between concentration and power; while in the meeting rooms the preferred color was blue and in the contact zones contrasting colors were chosen, like red and white, creating a different atmosphere from the rest of the space.

此项目的设计理念很好地表现出 TIBA
的企业文化。这是一个充满现代气息的
功能空间，极具西班牙风格。这是一个
充满朝气的公司，设计师创造出这样一
个工作环境：休闲区和会议室的天花板
可以升降，明亮的色调充满动感有助于
集中精力。

The design concept is the radical
cultural transmission of the company,
creating a modern yet functional space,
while following Spanish guidelines.
Because this is a young company, we
decide to give it an industrial ambiance,
with dropped ceiling just on top of
informal space and meeting rooms,
using bright colours to help people to
concentrate and be dynamic.

最重要的是，不论是公司的员工还是客户都能感受到设计的魅力以及公司先进、实用的服务。

Most important of all is that they—not only Tiba employees, but also its clients—feel by means of the architecture, the modern functional service that this company offers.

IMPLANT LOGYC

Implant Logyc 牙科诊所

设计公司：Antonio Sofan 建筑事务所
项目地点：美国弗吉尼亚州
项目面积：167.4 ㎡
摄影师：Todd Mason Halkin Photography

项目介绍 PROJECT DESCRIPTION

Implant Logyca 牙科诊所是美国最为先进的牙科、牙移植手术的诊所之一。它立志于为患者提供精益求精的牙科诊疗和最为人性化的治疗环境。

one of the most emotionally intense color + it stimulates
+ raise heartbeat and breathing + increases blood circulation
+ + and self-confidence + decreases lethargy and
+ helps ease chronic pain + may lower the frequen
+ wave lengths, causing patients to be less excited a
+ reduce pain + stimulates the automatic nervous a
circulatory systems + stimulates the liver + reduces infla
and swelling + supports the building of the blood

Orange, increases confidence and joy + decreases loneliness and fatigue + stimulates the lungs and thyroid gland to increase oxygen to the body + relieves cramping and convulsions in all parts of the body + relieves hiccups + builds lung and stimulate the respiratory system + could be used to treat Chronic Obstructive Pulmonary Disease (COPD), Emphysema, Chronic Bronchitis, Asthma and Tuberculosis

Blue, lowers negativity and stress + it is associated with thyroid, parathyroid, throat, mouth and lungs + acts as ... + it is useful in the treatment of skin disorders ... headaches + it is associated with wei... helps relax and calm muscles + it is associated with ... loss + relieves itching, skin irritations and inflammations ... helps with sleep disorders + rebuilds the skin when damaged by burns, scratches, infections or sores.

Green, is the easiest color on the eye and can improve vis + it is a calming and refreshing color + reduces claustroph bia, indecisiveness and anxiety + it is associated with the tuitary which is the master glands + it is useful for over healing, whether chronic or acute + helps fight infections helps in healing of sores, bruises and cuts. Stimulates t brain + stimulate the digestive system + helps in dissolvi blood clots

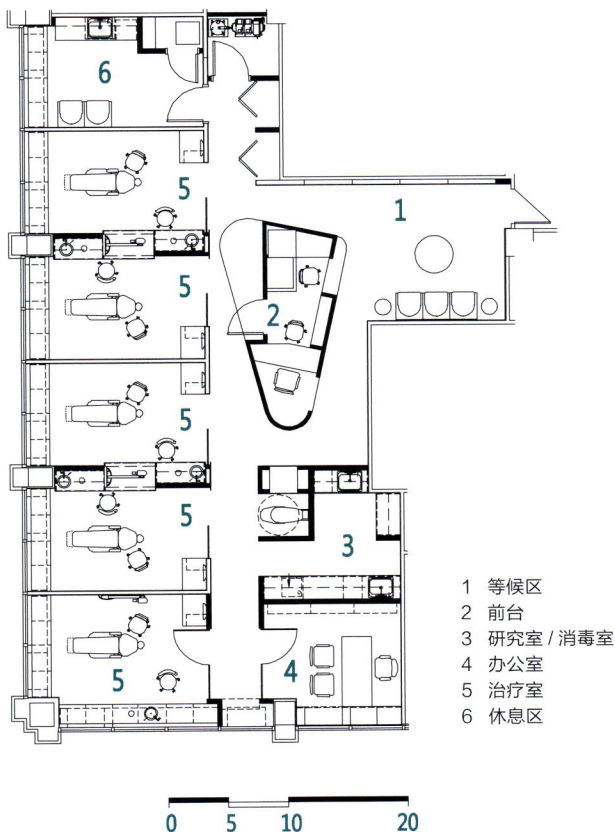

1 等候区
2 前台
3 研究室 / 消毒室
4 办公室
5 治疗室
6 休息区

0 5 10   20

"

这个牙科诊所位于美国弗吉尼亚州阿灵顿。诊所设计说明了在严谨的牙科医院背景中置入丰富色彩对治疗的有益作用，同时也说明了色彩能给患者和参观者带来一种积极而轻松的空间体验。一个带弧状角度的三角形体块上覆盖着厚薄不一、长宽不一的羊毛毡条，形成了一个 167 平方米的办公间，内墙为亮黄色，能源、行政和结账区分布在该空间的各个尽头。患者们从入口处穿过候诊区，到达前台，等待就诊；就诊完的患者安静地从另一边离开。这样的设计的意义主要在于能在紧张的空间中隔离出安静的手术室。

This dental office design in Arlington, Virginia is about making clean the therapeutic benefits of color as backdrops for the highly meticulous dental. It is also about articulating contrast in such a way that color provides positive distractions to patients and visitors as they experience through the space.

A curved triangular volume, cladding with natural wool-felt strips of different thicknesses and widths, dictates the flow of the 1800 square foot office, which is lined in bright yellow to encourage energy, admissions and checkout which are handled separately from inside at opposite ends. From the entry up to the waiting area patients walk up to the reception counter to proceed for treatment, then their signing out occurs inconspicuously on the other side. It was one of the main concerns to isolate acoustically the surgical site in such a reduced square footage.

沿着墙边设计了五个手术诊疗室，它们将开阔的视野引入室内。自然光线透过彩色夹层玻璃射进实验室和消毒室。每间诊室都有不同的颜色，使空间充满活力，并给患者健康带来正面的影响。每间房间以不同的材质巧妙地搭配各种色彩，让人们从较为压抑的灰白色走廊经过时能感受到轻松的视觉趣味。

> Five operatories laid out along the wall benefit from the urban and open views while allowing the natural light to be filtered through color-laminated glass into the lab and sterilization core. Each operatory is assigned a specific color that will help endorse vitality and generate a forceful impact on patients health. Skillful color-matching materials inside each room unleash interesting optical illusions when walking in and out from the main corridor which displays more muffled colors like white and gray.

迪斯尼专卖店总部

设计公司：Clive Wilkinson 建筑公司
项目地点：美国加州
项目面积：7530 ㎡
摄影师：Benny Chan/Fotoworks

项目介绍 PROJECT DESCRIPTION

迪斯尼专卖店（Disney Store）是专门设计销售迪斯尼各类产品的全球专卖店连锁，第一家迪斯尼专卖店于 1987 年在美国加州格伦代尔开业。

美国迪斯尼专卖店总部办公室是由 1927 年建的旧仓库改造而成的。设计师在 7530 平方米的两层楼中，设计了一个既实用有趣又突出迪斯尼形象，同时可容纳 230 名员工的开放、灵活且交流方便的工作环境。该项目在建筑和装饰上巧妙地运用了彩色模块和灵活的装配系统。

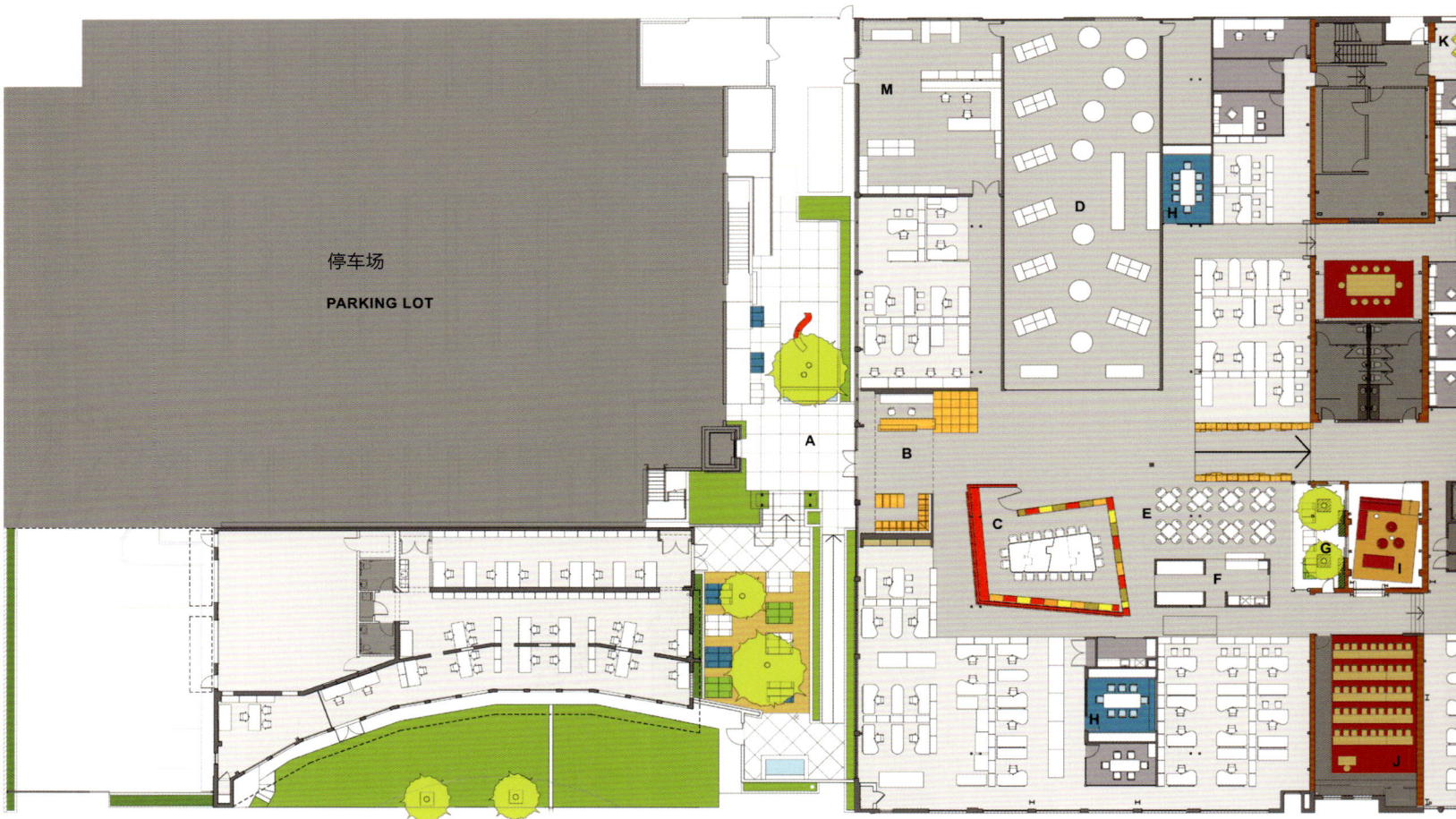

停车场

PARKING LOT

A – exterior courtyard 室外庭院

B – reception 前台

C – block conference room 砖房会议室

D – mock store 模型商店

E – cafeteria 自助餐厅

F – pantry 储藏室

G – exterior lightwell 室外采光井

H – workrooms 工作区

I – brainstorming room 头脑风暴室

J – training room 培训室

K – informal lounge 休息室

L – honeycomb conference room 蜂巢会议室

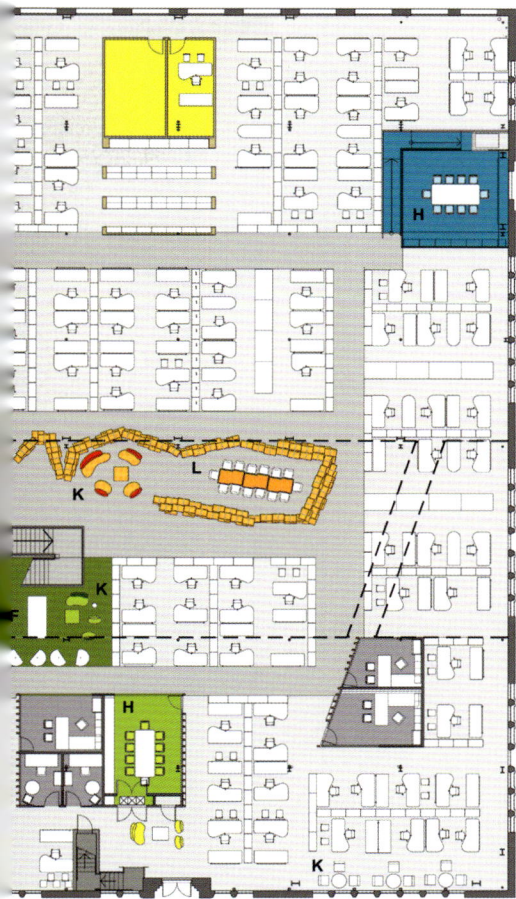

M- mailroom 邮寄室

The Disney Store Headquaters's West Coast headquarters was built in historically significant Royal Laundry Building (a former laundry built in 1927). The design solution for the 7,525sq.m. space evolved from the desire to create a functional yet playful environment that befits the Disney image. Equally significant was the company's desire for an open, flexible and collaborative work environment for 230 employees.

"

空间的大部分区域被两间主会议室占据，其中第一间是两边用可以拆卸的泡沫砖墙组成的"砖房会议室"，在开公司大会时，这些红、橘、黄、赭的砖可以拆开充当200个凳子。这些彩色砖块是从现有砖墙中得到的灵感。

另一个会议室是"蜂巢会议室"，位于仓库的中心位置，可以灵活展示迪斯尼的玩具样品，巧妙装饰了空间，也带来了儿童玩具设计的灵感。

除了一个室内的景观区外，在公司入口处还有一个室外庭院，里面有修剪成"米老鼠耳朵"造型的植物。

The two major portions of the building are anchored by two main conference rooms. The first, known as the "Block Conference Room" is formed on two sides by removable foam block walls. When these foam modules are disassembled for 200 person company-wide meetings, they become the seating system. The red, orange, yellow and ochre color palette was inspired by the existing brick wall colors.

The second main conference room, formed by a unique modular honeycomb structure, is located in the Atrium portion of the building. Originally conceived as a flexible means of managing the Disney sample product display, it became the centerpiece of the space.

In addition to providing an internal landscaped courtyard and new skylights throughout, the building connects occupants to the exterior with a new landscaped courtyard at the front entrance, which includes an ivy topiary of "Mickey Mouse ears".

TBWA CHIAT DAY

TBWA Chiat/Day 洛杉矶办公室

设计公司：Clive Wilkinson 建筑公司
项目地点：美国加州
项目面积：11 160 ㎡
摄影师：Benny Chan/Fotoworks

项目介绍 PROJECT DESCRIPTION

TBWA Chiat/Day 是 TBWA 腾迈公司的美国分部，在美国共有四个办公点。作为老资格的广告品牌，Chiat/Day 曾为苹果公司等大品牌制作过著名广告。1994 年，试图降低运营成本、提高工作效率的 Chiat/Day 又出惊人之举，把原来的办公室改成仓库，让员工拎着笔记本、手机回家，实行虚拟办公！但事与愿违，虚拟办公导致了工作效率更低、大批员工离职。1995 年，Chiat/Day 被 Omnicom 收购，并与 Omnicom 1993 年时收购的 TBWA 合并，形成现在的 TBWA Chiat/Day。

该项目在大仓库中建造了一个 "广告城"。这个项目通过高层建筑、绿色园林和不规则的 "天际线"，改善了这个迷你城的环境。门房由雕刻的金属层包裹，形状像一个胶囊，形成一座通向主入口的标志性建筑。前台、走廊展示了广告公司的作品。门房与主建筑区之间由两座管状人行天桥相连，一座通向一层，一座通向二层。

The space is the creation of an "advertising city" constructed inside a large warehouse. The program offered the chance of developing this small city environment with multiple levels, green park space and an irregular "skyline". A sculptural metal clad "gatehouse" pavilion was proposed to accommodate the agency's main entrance and to provide an identifying landmark. The warehouse site offers a capsulized entry in the form of a gatehouse. This structureaccommodates a reception area and gallery for displaying the agency's work, and is connected to the main warehouse by two pedestrian bridges like long tubes, one leading to the ground floor and the other one to the second level inside.

空间采用了传统的城市规划，包括主街道、居民区、公园、胡同和街面设施，它们将原本粗糙的厂房变得更加人性化。周边的临时空间用布帐篷搭成，便于移动，并提醒人们此地正在施工。

Traditional urban planning concepts of Downtown, Main Street, neighborhoods, parks, alleys, and street furniture all became planning elements necessary to humanize the raw industrial space. Temporary structures were best formed in fabric tents, allowing easy replacement and reminding people of shrouded buildings under construction.

底层的中间有一条主道贯穿，各个部门和功能区由主道、桥和通往中间层的斜坡相连。三层楼高的建筑使整个空间充满城市气息。主道两旁是外形像洞穴的"悬崖屋"。"悬崖屋"是用钢构造建成的，以减少建筑之间的间隙。木质结构成本较低，用于大部分房间的建造。会议室由三个定制的集装箱临时组成。

The various departments and service facilities are to be connected through a "Main Street" which bisects the ground floor, and bridges and ramps that connect mezzanines. At three stories high, these structures lend a sense of urban scale to the whole agency. The design of the "Cliff Dwellings" is either side of "Main Street" where creative teams operate in mechanistic "cave" structures .

巴洛塔慈善总部办公室

设计公司：Clive Wilkinson Architects 设计公司
项目地点：美国洛杉矶
项目面积：4371 ㎡
摄影师：Benny Chan，Fotoworks

项目介绍 PROJECT DESCRIPTION

巴洛塔慈善机构是美国慈善事业的成功典范，非常善于激发公民捐款的积极性。它曾策划新颖的"抗艾滋病自行车拉力赛"、"抗艾滋病非洲长跑赛"、援助自杀者"走出黑暗"活动等，成效显著。它的慈善理念和筹款方式被英美等无数国家效仿，目前它已为重大事件筹集了数千万美元的善款。

通过一个印有花纹的大遮阳棚进入办公室，首先进入视线的是一个以戴马克松世界地图为原形创作的小岛状前台。走过前台，入口处设计了一个深蓝色的两端未封闭的集装箱，使大街与办公室主要区域相连。六个橙色集装箱堆叠后构成了三层行政楼。行政楼背后是两层高的木质会议室、多媒体工作室以及员工工作区。设计旨在达到舒适实用目的的同时坚持绿色环保原则。

Entering the building through a large screen-printed sunshade, the reception area features an island desk modeled on Buckminster Fullers' Dymaxion world map. From this area, a dark blue open-ended shipping container forms a portal to the main volume of the building and onto the main street, leading on to the square with its executive tower, a 3-high six-pack of orange containers. The backdrop to the square is a 2-story gallery of rough wood framing containing meeting rooms, audio and video recording studios, and support staff working areas. The project has achieved the targets of comfort and utility, as well as supporting a responsible green approach to resource efficiency.

The bitterest tears shed over graves
are for words left unsaid
and deeds left undone.

HARRIET BEECHER STOWE

"

Color on the project is very deliberately used to choreograph views and to distinguish public and neighborhood zones. A dark blue open ended shipping container creates a deliberate transitional experience from the bright entry into the interior landscape beyond, framing ones initial view into the dramatic whitetent landscape. The palette was developed to subtly vary the views and enhance the sense of depth and composition of neighborhood forms without confusing the clarity of the tents. Set against this saturated palette is a variety of more muted earthy tones used on the floor of the tent neighborhoods. In combination with the exposed concrete in the circulation zone, this ground plane is reminiscent of the actual ground present in the mobile tent cities that the charity creates.

办公室的整体色调经过了精心配置，色彩跳跃性强，公共区域和邻近区域都有明显的区分。深蓝色的集装箱是亮色入口和内部的过渡区。集装箱上的白色帐篷和蓝色箱体形成了视觉对比。这样的色彩增加了场地设计的立体感。咖啡室和主楼给一大片白色和蓝色的区域添加上了惹眼的橙色。光泽涂料更增添了艳丽的视觉效果，而帐篷地面柔和的土色又中和了这种艳丽。过道上，裸露的混凝土质感让人感觉仿佛置身户外，这里放置着用于慈善救济的移动帐篷。

SEAT YELLOW PAGES

Seat 黄页广告公司

设计公司：Iosa Ghini Associati 设计公司
项目地点：英国伦敦
项目面积：26 000 ㎡
摄影师：Roberto Centamore、Juergen Eheim

项目介绍 PROJECT DESCRIPTION

该公司隶属于号称"意大利电信"的 Seat PG，在意大利有着深厚根基，网络覆盖 180 家多媒体运营商。除了传统的视觉广告，该公司还发展了多平台广告媒介、多媒体平台、互联网广告等信息交流平台，拥有数千万用户。

堪称世界最大的多媒体黄页广告公司之一的 Seat 黄页广告公司的行政总部，坐落在都林边界上一块 200 万平方米的地方，此处有许多废弃的大型工厂。

Spina 3, an area of more than 2 million square meters at the edge of Turin and nearly completely comprised of large historic factories in a state of abandon, has become the new home for the offices and executive headquarters of one of the world's largest multimedia directory advertising companies.

项目包括 6 栋四层大楼，设计利用了大楼原有的设施。前台和公司总部被选在 19 世纪的一座大楼中，这座建筑分为五个不同的部分，行政办公区、教室、公司陈列馆、礼堂以及酒吧餐厅，每个部分都配备先进的技术设备。

The project includes six new four-story buildings and reuses the original building formerly the assembly site of airplanes and engines for submarines and blimps. The main 19th century building was chosen as the site of the reception and the "home" of the company – prior to the move, Seat was spread out into five separate structures – connoted by a system of advanced technological infrastructure, operating and executive offices, classrooms, the Seat museum, an auditorium and lounge bar annexed to the restaurant.

内部结构的设计同时体现了这家商业传媒巨头和建筑公司的设计思路，另外也在建造过程中征求了施工方的建议，最终设计出一个集功能性、舒适性为一体的工作空间。

Redesign of the interiors reflects the common design process made by a scientific group of personalities in the world of business and communication and an architecture office, which also incorporated the views of the workers in a process of "participatory design" to create a functional work place, which would also meet the need of wellness and comfort.

LEGAL OFFICE

Legal Office 法务办公室

设计公司：Creneau International
项目地点：比利时
摄影师：Philippe Van Gelooven

项目介绍 PROJECT DESCRIPTION

Creneau International 相信办公设计没有戒律，他们喜欢打破常规，维护创意的权利。在他们看来，即使是法务办公室，也应当表现出自己的设计理念。

接待室的墙上有两个艺术装饰画框，上面有创办者和他儿子的激光签名，表达着他们对来访者的诚挚欢迎。主会议室旁边有一部用电缆线固定着的没有扶手的楼梯。定做的木头桌面中和了空间整体的工业感。私人会议室中只有一道秘密的楼梯通往主会议室，保证了空间的私密性。会议室有一面模仿书架做成的双层墙，便于隔音。古典的木线装饰在水泥天花板的四个角上，让这里充满视觉对比和独特创意。

In the waiting room, two artworks featuring the lasered signatures of the founder and his son function as a personal welcome to the visitors.

Over to the main meeting room, via a "naked" staircase that appears to be kept in place with steel cables. Wooden table tops introduce warmth into the industrial frame of the office. Piece of resistance in the building is a private meeting room that can only be reached through a secret stairway in the main meeting room. Featuring a wall of books (fake) that doubles as an acoustic screen, and classic cornice mouldings framing the concrete ceiling, this room exudes contrast and creativity.

PAPSA

海沃氏公司墨西哥城办公室

设计公司：Space 设计公司
项目地点：México City
项目面积：800 ㎡
摄影师：Santiago Barreiro

项目介绍 PROJECT DESCRIPTION

总部位于美国密歇根州的海沃氏公司是一家全球性的办公家具制造商。一直以来，海沃氏公司都占据着高端办公家具的市场。据海沃氏公司 2008 年全球可持续发展报告显示，该公司在 120 个国家设有分支机构，在 10 个国家拥有超过 20 个独资的制造厂房以及 600 余个分销商。

水、大理石、竹子是给到访者带来新鲜感觉的三个元素，使其领略参观的第一个集合点："对抗室"。人们从这一空间设计中可以体会到可持续发展观念，了解项目的发展以及能源与环境设计先锋奖的信息，增强环境意识和责任感。

Water, marble and bamboo are three elements that give the visitor a fresh welcome, inviting him to the first meeting pcint in the tour: the "war room". To raise awareness and promote responsibility towards our environment and Mexico, the user receives in this space information on sustainability, the development of this project and the LEED-CI certification process.

> 这个项目的核心区域在会议和展示区，这里由一个会议室和一个展示厅组成，花格窗用透明的配线架做成。在这个区域周围，开放式工作区被分成三个不同的区域。每个区域根据功能设计，由家具隔开。为了改善采光，有的用比较矮的挡板，有的放置了透明的隔墙。办公室的四周设计了灵活多变的走廊，平衡了空间感。主道的尽头是经理的私人办公室，同样也设计了透明的隔墙。

The heart of the project is made up by the meeting and exhibit areas which consist of a boardroom and an exhibit room confined by an apparent MDF lattice window. Around this core, an open work area divided into three different areas arises. Each one of these areas is resolved with furniture specifically designed for its role, whether they are low room dividers or elements of transparency strategically placed to let light in as much as possible. On the perimeter of the office, a corridor generates lively tours, which intentionally causes a spatial equity for users. At the end of main circuit we find the directors' private offices, which retain the same language of transparency.

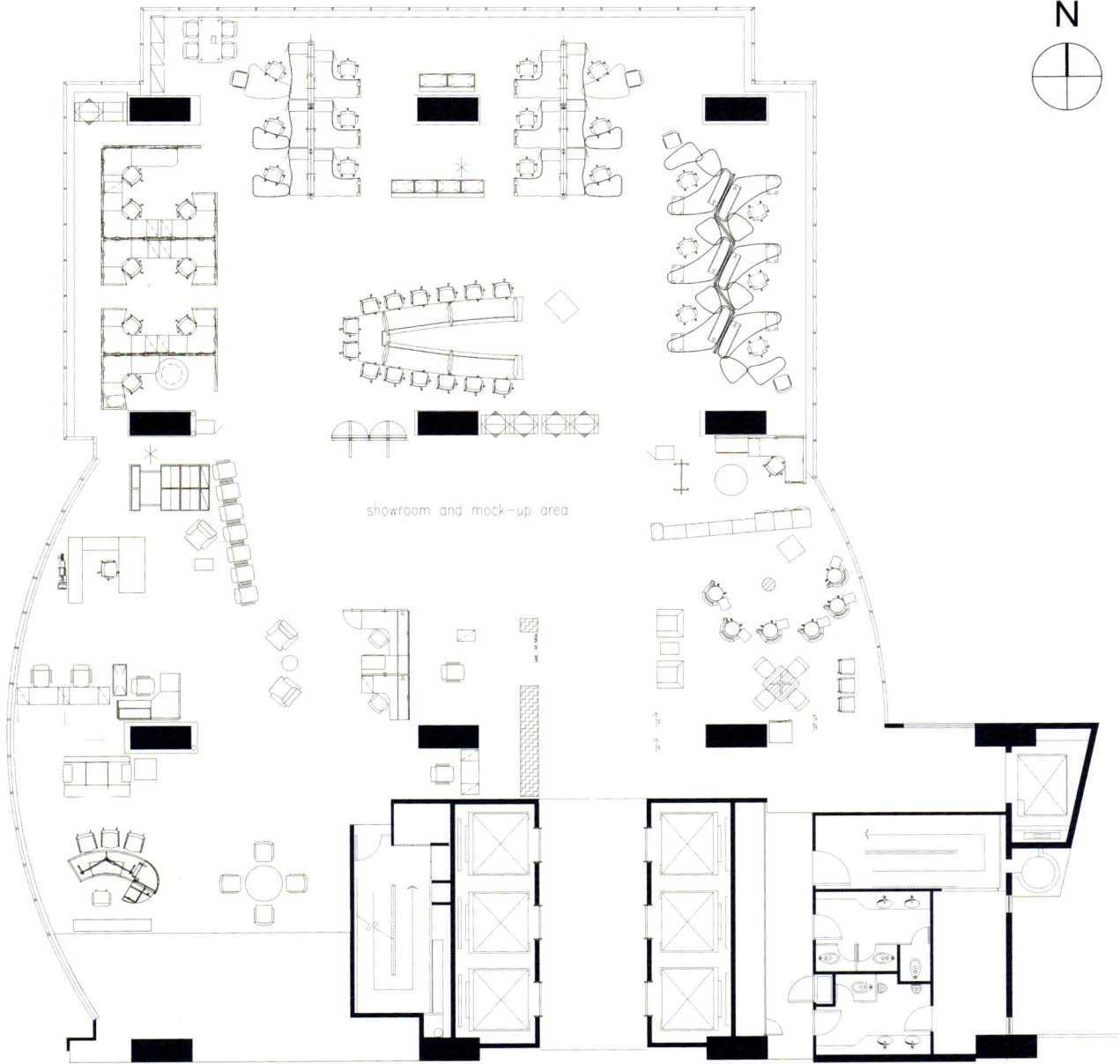

showroom and mock-up area

N

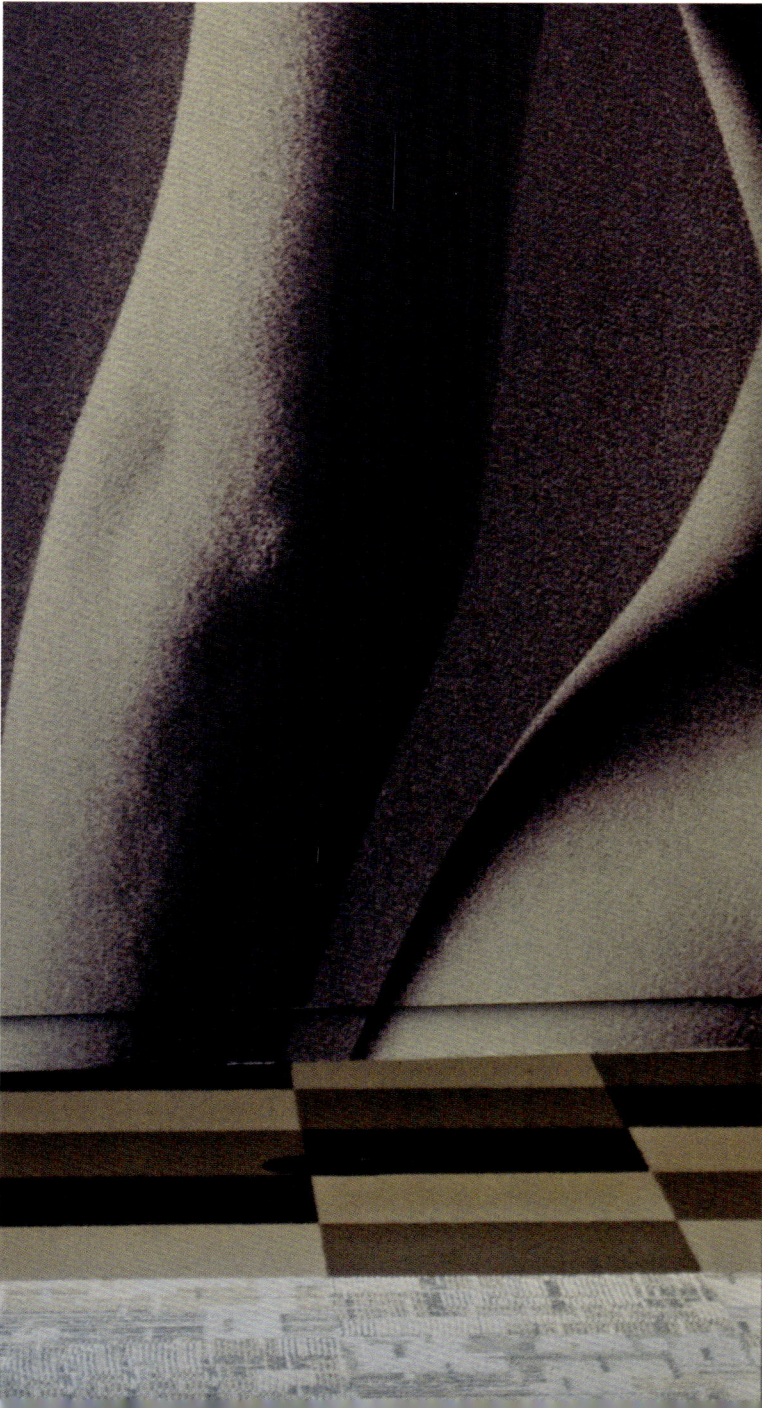

项目的基本结构围绕可持续发展而设计，运用整体手法体现在实质、观念和形式上。从活动地板到家具，使用的材料均为绿色环保材料。

The project's fundamental concept revolves around environmental sustainability in substance, form and conviction, in a holistic manner. The materials used in the project are sustainable-green, ranging from the false floor to the furniture.

ONESIZE

Onesize 公司

设计公司：Origins 建筑事务所
建筑公司：Kne+
项目地点：荷兰
项目面积：300 ㎡
摄影师：Stijn Poelstra

项目介绍 PROJECT DESCRIPTION

Onesize 公司是荷兰一家创意工作室，于 2001 年成立。该公司主要专注于制图、动画、视觉特效、电影电视道具制作等创意设计。经过多年的发展，公司已与许多著名公司合作。

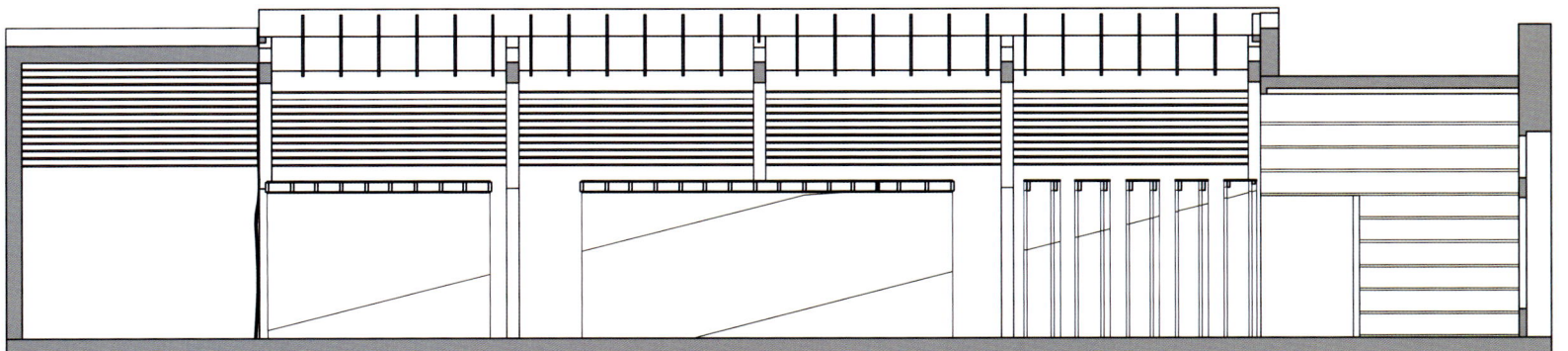

他们使用一般用做地毯底衬的低等杉木做材料。除了成本因素，高质量的细节设计和低档次的材料搭配有很好的突出效果。这种强烈对比和现有建筑与中心雕塑的对比相呼应：木质材料与混凝土，细节与材料，明暗对比。

They used low grade spruce which is usually used under carpeting. Besides the cost issue, they strongly believed that the juxtaposition of high definition detailing and a low grade material would make both stand out better. This contrast is also echoed in the relation between the existing building and the central sculptural shape: Wood and concrete, detail and material, dark and light.

> 设计最开始是复杂的形状，但完工后显得更简单些。考虑到可持续性的因素，设计也关注环保影响，制作过程中木材的损耗比较少。设计师充分考虑到室内的音响效果、照明以及防火设备，设计出比较合适的方案。项目最重要的是设计和外观都要适合客户。

They started out with more complex shapes, but the simpler it became the better the result was. Besides, their office specializes in sustainable building, so they were also keeping an eye on the environmental impact. By doing so we actually came up with a sculptural volume that hardly has any saw losses in the making. They made a few key decisions on the materialization of the interior spaces so that acoustics, lighting, fireproofing etc. were at handled properly. The most important result is that the interior really fits the client, both in terms of program and in appearance.

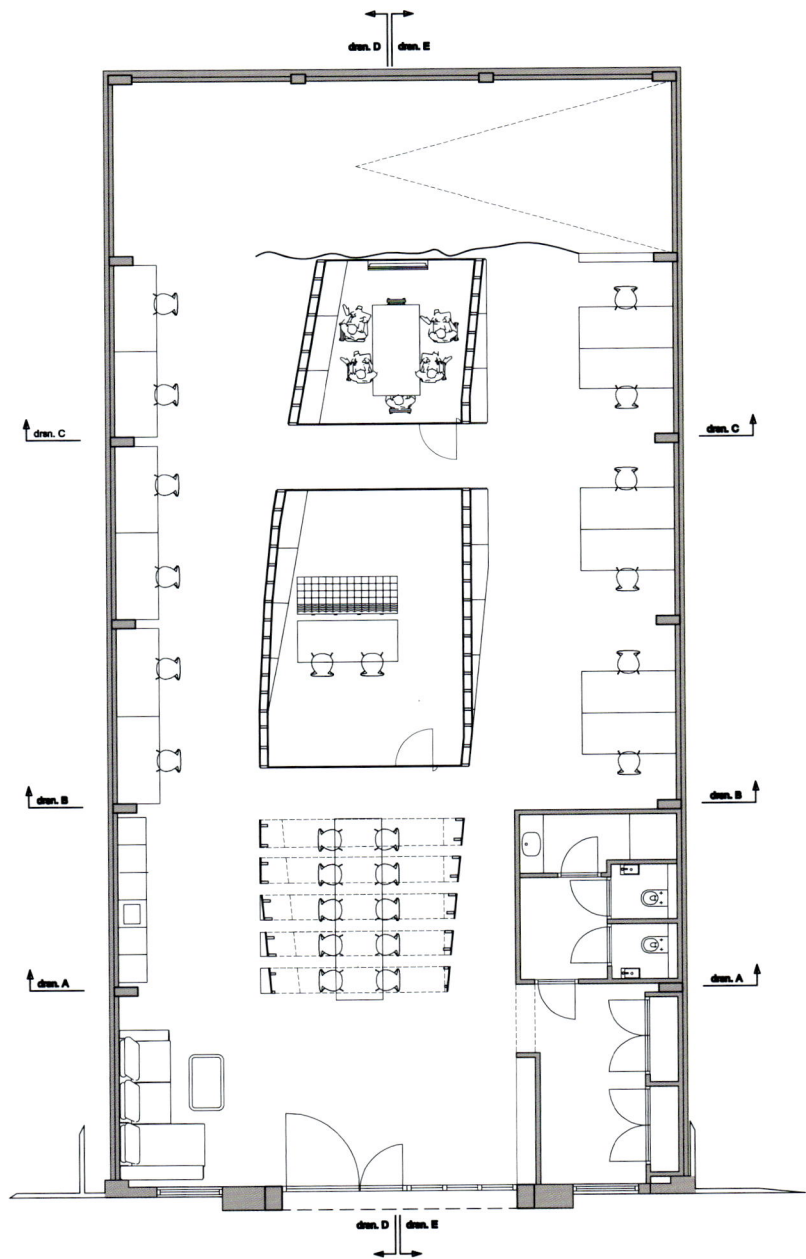

LG DESIGN CENTRE

LG 欧洲设计中心

设计师：Shaun Fernandes、Simon Jordan、Go Sugimoto
设计公司：Jump 工作室
项目地点：英国伦敦
项目面积：600 ㎡
摄影师：Gareth Gardner

项目介绍 PROJECT DESCRIPTION

LG 集团是消费性电子产品与移动通讯设备的全球领先企业。其设计产品涉及日用电子产品、手机及其配件、宽频彩电与音响系统以及白色家电等广泛领域。

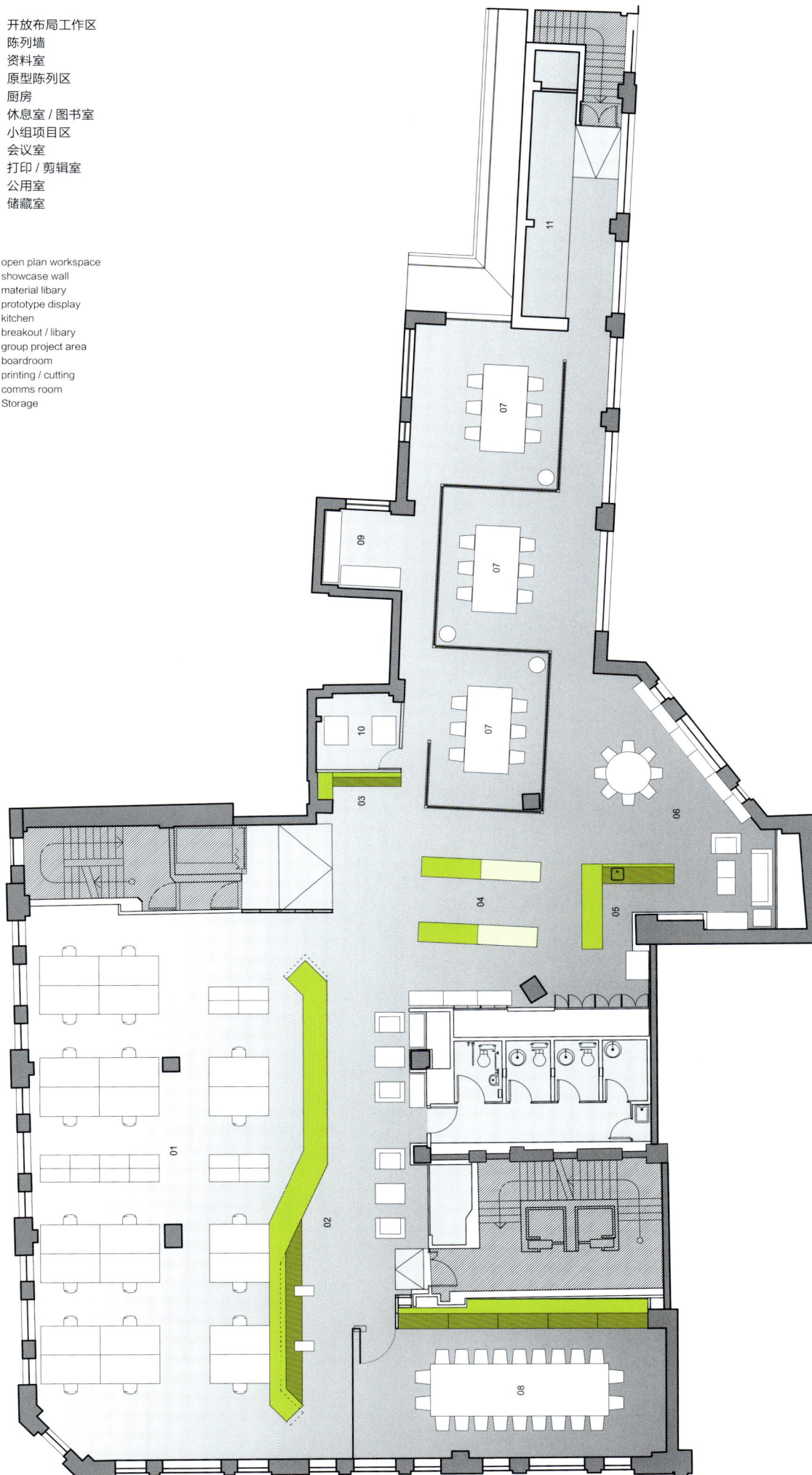

"

最近，一个全新的 LG 欧洲总部在伦敦建成，欧洲设计工作室是 LG 电子产品公司的最重要的设计中心之一。这里有 20 位来自欧洲的工业和界面设计师。清爽的配色、充满设计感的桌椅成为这个宽阔空间的焦点。藏书室中的阅读台方便而现代。

A new European Design HQ for LG has recently completed in London. The European Design Studio of LG is one of the most important design centres for the electronic goods company and houses a team of 20 dedicated industrial and interface designers, drawn from around Europe. Fresh and cool colors, with the tables and chairs full of design feeling, become the focus of spacious office. The reading sets in the library are convenient and modern.

"建立新设计中心是希望创造一些包含大面积藏书室的创意空间，为我们的产品设计提供技术支持。欧洲和其他地区的市场之间存在着巨大的科技文化差异，而 LG 的设计目的就是要密切契合欧洲客户的心理，从而应对这种差异。"LG 欧洲电子总裁兼 CEO 说。

"Our new European Design Centre has been established to provide our design team with the optimum creative environment, which includes extensive libraries and leading technologies to aid in the design process. There are significant cultural and technical differences between Europe and other markets. As a result, for LG to grow its market presence in Europe, it must invest in and be sensitive to the consumers' requirements, while responding to technical and cultural differences," said president and CEO of LG Electronics Europe.

BLOOMBERG

彭博商业资讯总部办公室

设计公司：Jump Stuidos
项目地点：美国

项目介绍 PROJECT DESCRIPTION

Bloomberg 商业传媒公司是 Bloomberg 集团旗下的分公司。该公司是金融软件、传媒和信息的行业巨头。
商业传媒公司的独立，标志着该公司将独立运作商业新闻报道、广告媒体等业务，同时也优化了企业的
资源配置，提高了工作效率。

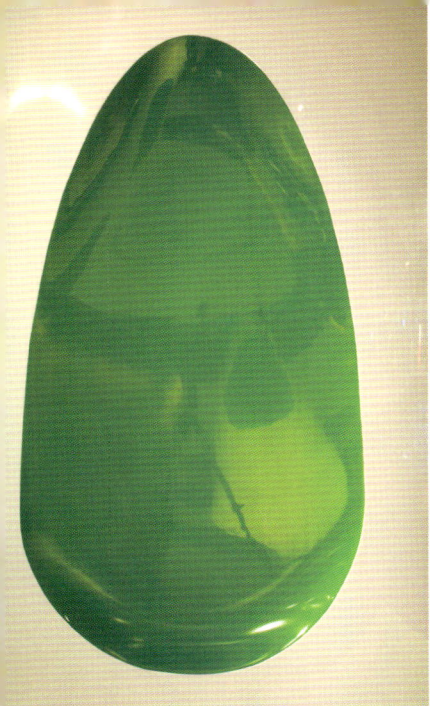

Bloomberg 城市总部设计的亮点是每层楼的集会区，其中间的座位形状各异，树的主题贯穿其中。最底层的会议区扮演了树桩，这种设计考虑到底层工作人员更易受来访者的打扰；中间层的座位变成了枝权，造型大方，给人印象深刻；最高层的软座就像树叶，绿色和谐。这样的设计将整座大楼的设计风格和谐统一起来。会议区的外墙包裹着透明玻璃，从室外看就可突出公司形象。办公区统一布局在每层会议区的左边，由一层玻璃隔开，来访者一眼便可看到办公室内的情况。整层楼的布局简洁，可变性强，旁边的自由办公点设计现代实用。

The project, across three floors of Bloomberg's City headquarters, is inspired by a tree. The lowest floor features meeting spaces created in the trunks, and the middle floor incorporates seating in the "branches", while the top level includes soft seatings which resemble the foliage. The same areas on each floor clad with transparent glass through the three floors transfer a harmonious and natural concept while indicating identity of the company. Workplace is designed on the left of this area and shut off by a glass wall, besides modern, flexible and practical free working points.

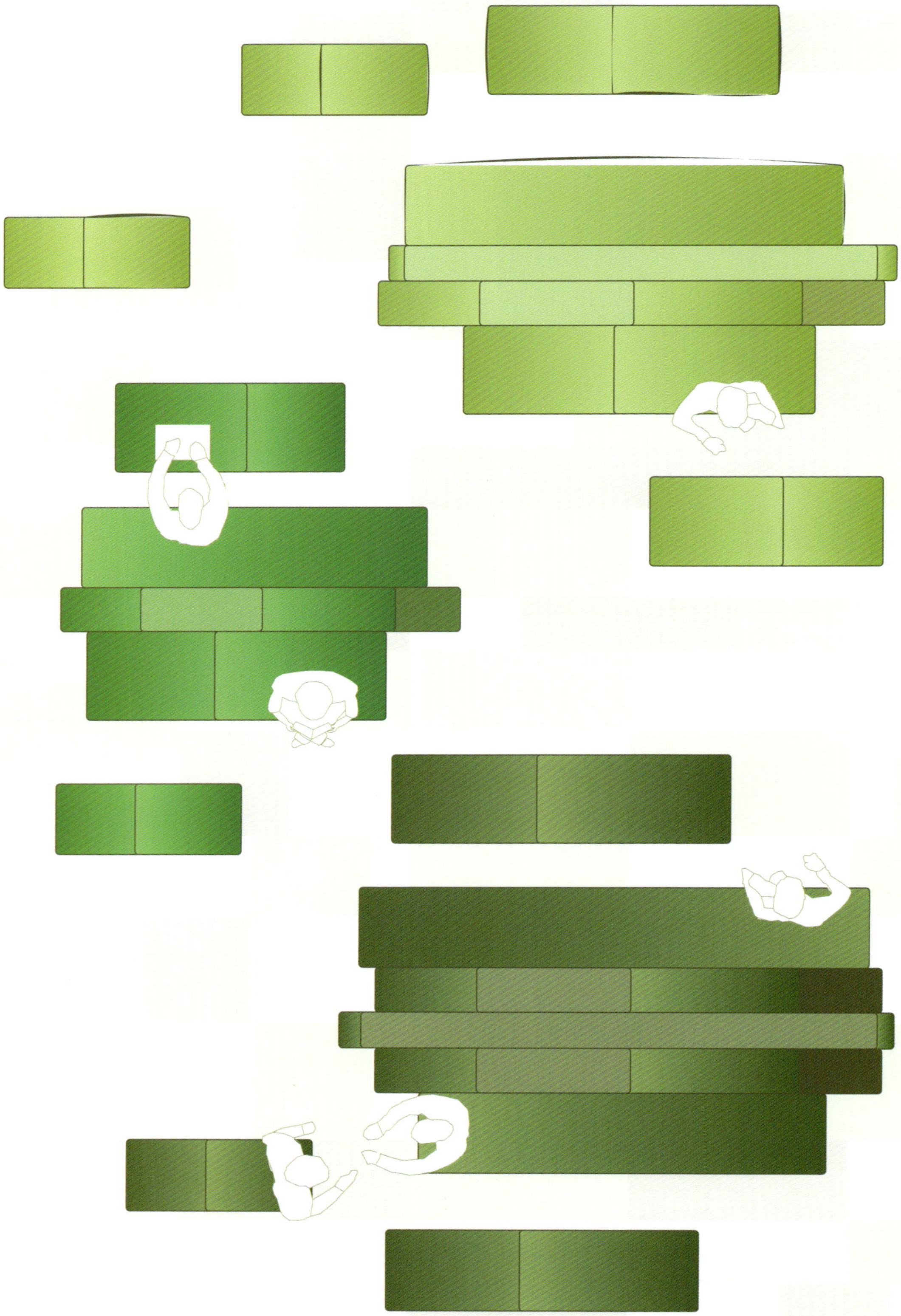

BLOOMBERG

244

GALCIT

美国加州理工学院航空航天实验室

设计公司：John Friedman Alice Kimm 建筑公司
项目地点：美国加州
项目面积：3070 ㎡
摄影师：Benny Chan/Fotoworks

项目介绍 PROJECT DESCRIPTION

美国加州理工学院航空航天实验室（GALCIT），是闻名世界的顶级航空科技实验室，为全世界的宇宙航天及空间探索事业做出过重大贡献。

> 在考虑整个实验室的多功能特性时，设计师想到了"流体"的概念。而实验室大部分工作都是用于研究流体，观察固体、流体、气体在不同压力下的变换。从天花板到墙纸的设计，设计师把实验室变成一个灰白的容器，在这里人们可以从墙面、柱子和空隙中研究流体的图案。实验室的整体设计是一个统一的设计理念，即一个容纳科研实验室、教室、研讨室、办公室的功能强大的空间。

After thinking about the multifunction union of the whole laboratory, we ultimately derived much of our formal language from the notion of FLOW. Almost all GALCIT research involves the study of flow – understanding how solids, liquids, and gases behave under differential pressures. We also began to think about it as an architectural wind tunnel – a relatively neutral container into which we could drop objects – in the form of new ceilings and wrappers, and metaphorically study their flow patterns as they interface with existing walls, columns, and voids. By gathering together all of the spaces under this rubric, we attained a very specific expression for each space, while maintaining continuity between them. We also created highly functional spaces, fulfilling every nitty-gritty requirement of the academic laboratories, classrooms, conference facilities, and offices that comprise the program.

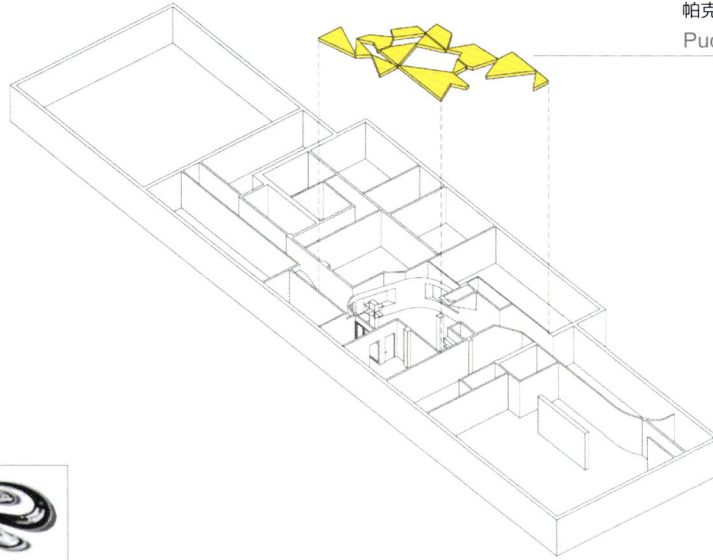

帕克特会议展览室
Puckett Conference and Display

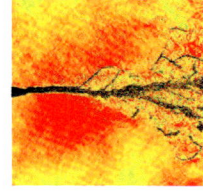

Cann Laboratory instructional space
卡恩实验室的培训室

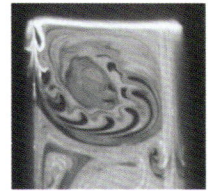

互动休息室 interactive Lounge

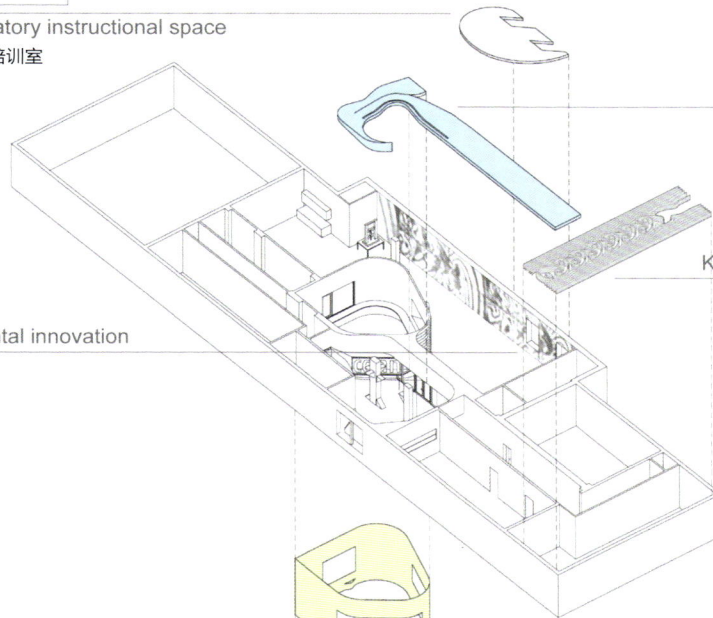

进行创新实验的卡恩实验室
Cann Laboratory of experimental innovation

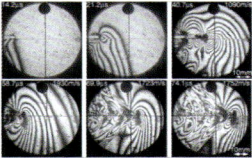

卡曼会议室和加州理工学院航空航天实验室的档案室
Karman Conference Room and GALCIT Archives

double height wrapper 双层高包装墙纸

lobby 门廊

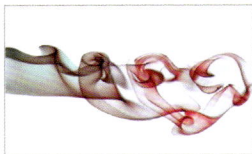

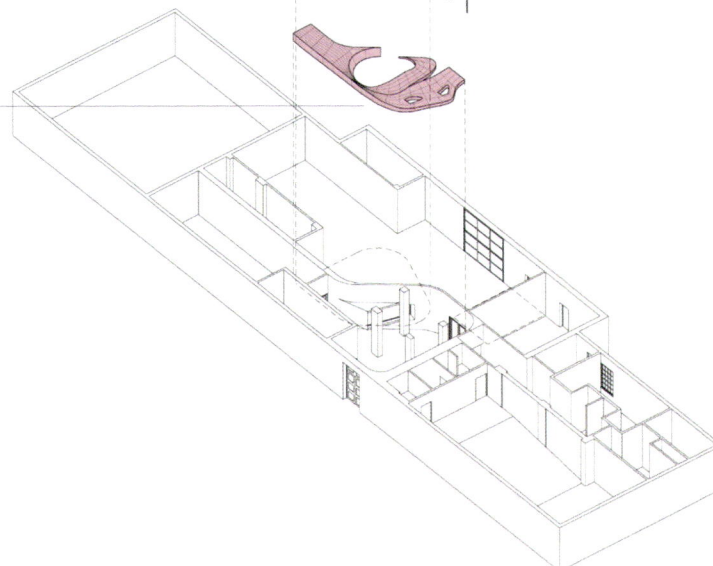

大厅：大厅是入口的重点，是建筑的首个公共集会和交流中心。一入大厅，带流线花纹的塑料吊顶即映入眼帘。波浪是根据光源设计的，如同光源搅动起水面泛起的涟漪。

一层的太空结构实验室：这个实验室包括一个小展厅，里面展览了教授的研究成果。实验室里面的双层区域用于大型试验。

二层的互动休息区：波浪形的吊顶环绕立柱结构，在外墙和二层实验室之间建立一个公共空间。

二层的"卡门涡流"会议室和陈列室：宽阔的会议室和展览设施，吸音天花板采用卡门涡流式造型。玻璃、亚力克、钢质桌既可以作为展览品，也可以作为会议桌使用。

三层楼的会议室和陈列墙：设计师利用几何造形把一块原本封闭狭窄的区域转换成会议室和流动展示墙。

FCB 国际办公室

设计公司：Clive Wilkinson 建筑公司
项目面积：9300 ㎡
项目地点：美国加州
摄影师：Benny Chan – Fotoworks

项目介绍 PROJECT DESCRIPTION

1963 年，作为世界广告史上元老之一的 FCB 在纽约证券交易所上市，成为世界上第三个上市的广告公司。如今，FCB 已经成长为一个强大的全球办公网络，在 110 个国家拥有超过 190 个办公地点。在分工协作、资源共享、传媒互动，销售产品的同时建立品牌资产的创意，是企业的最终目标。

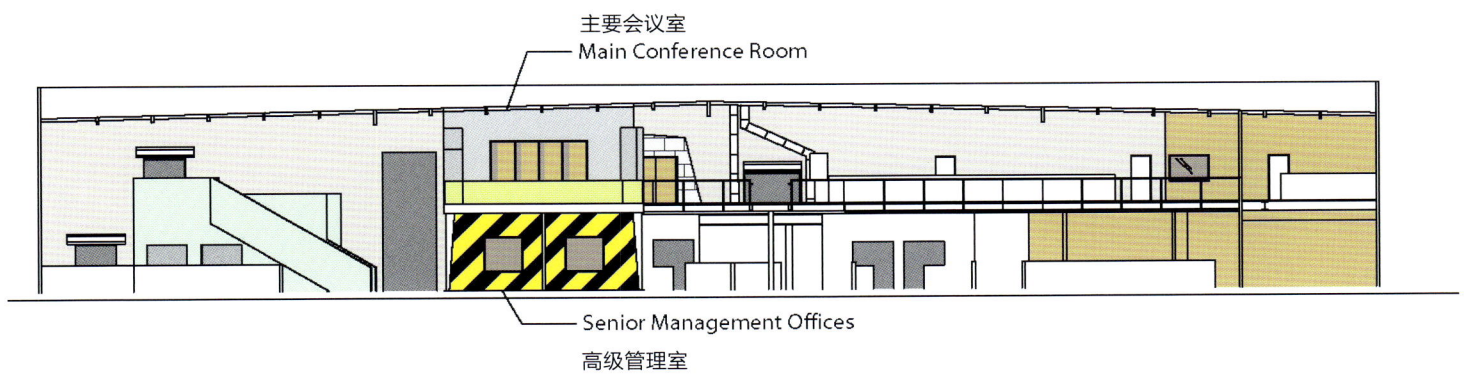

主要会议室
Main Conference Room

Senior Management Offices
高级管理室

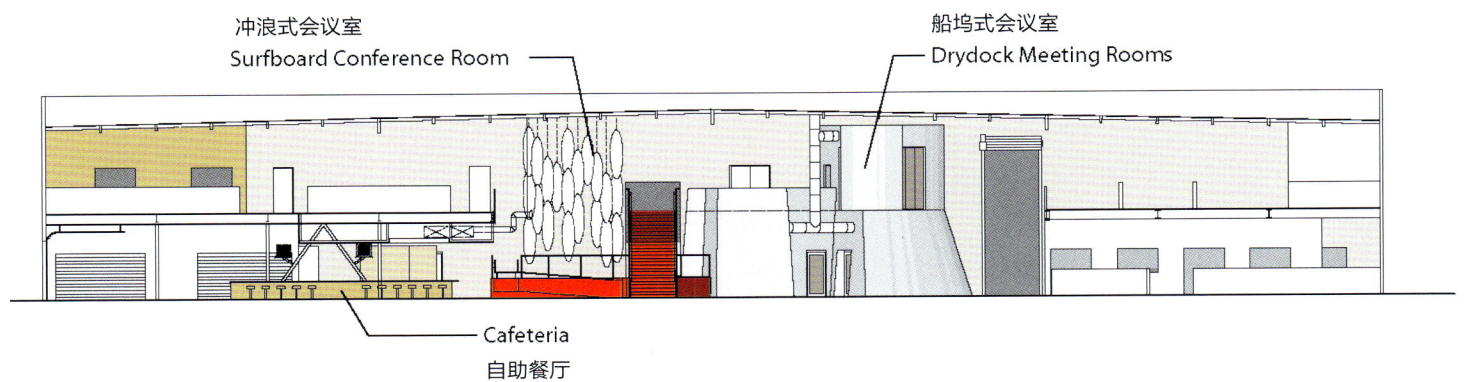

冲浪式会议室
Surfboard Conference Room

船坞式会议室
Drydock Meeting Rooms

Cafeteria
自助餐厅

FCB 客户要求办公室具有前卫的创意。他们希望打破旧式思维，激发员工的创造力。办公室具备开放、协作、高效运作的特点，通过打开传统的密闭办公空间来增加员工的视觉交流。

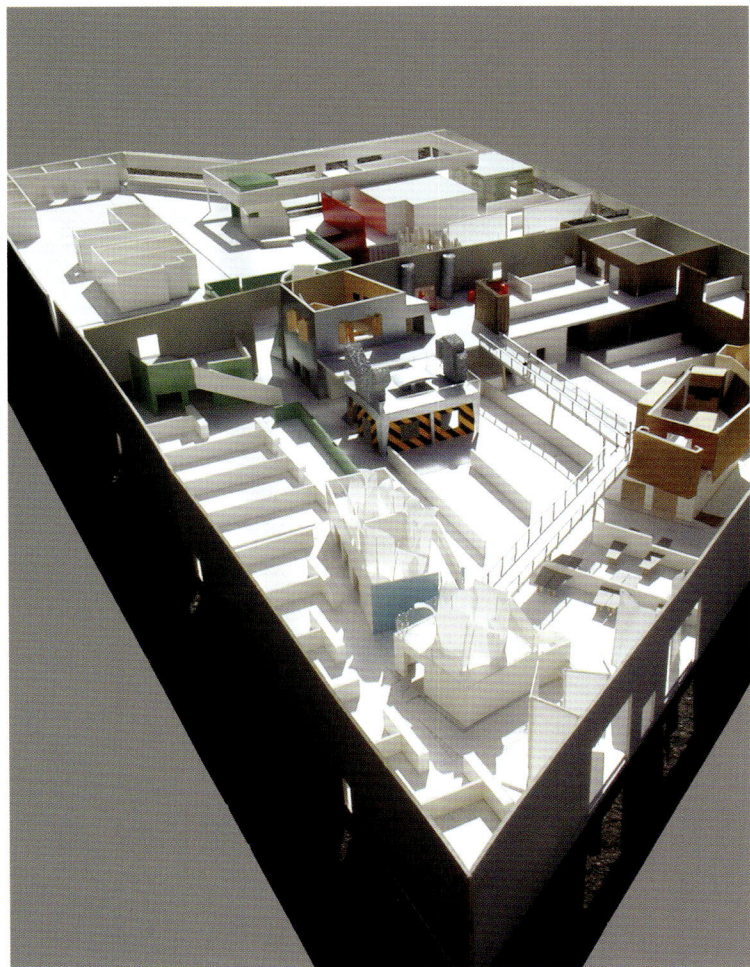

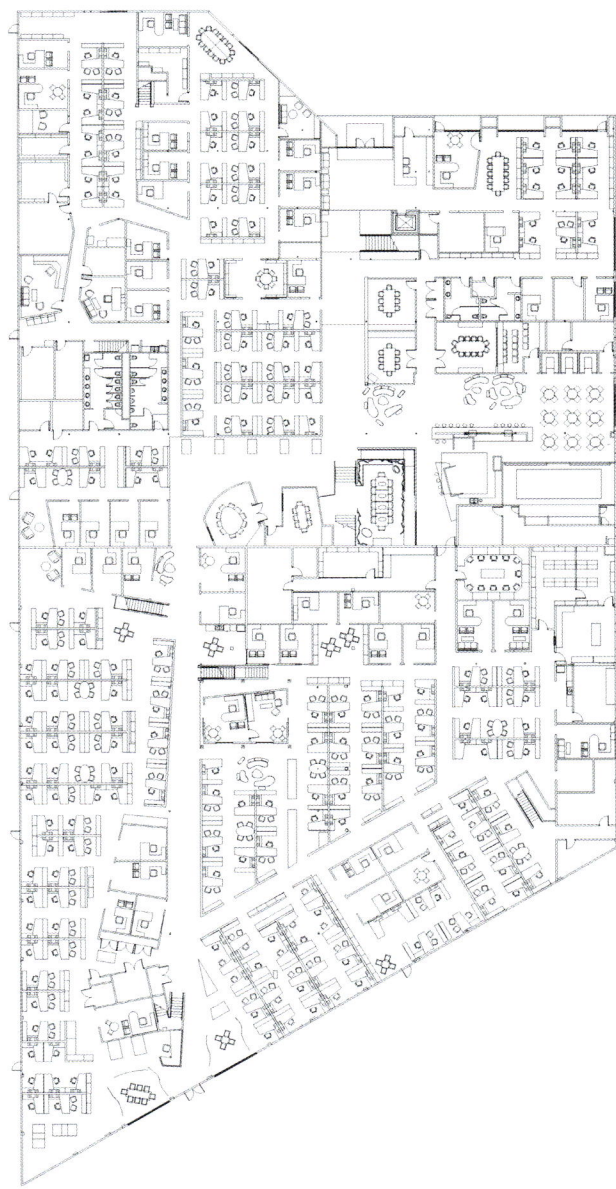

0
10
20
40

"

With FCB's goal of reinventing their agency as a cutting edge creative house, they wanted to lay down the groundwork by challenging preconceptions about how an office space could work and how it could serve to inspire their staff into a new way of embracing creativity. This project allowed us to systematically peel away old status-based planning ideas, opening up private offices, and improving visual contacts in work areas.

Continuing up the main red staircase, which penetrates an existing concrete wall, you will arrive at the mezzanine overlooking the creative warehouse area.

走上红色的楼梯，来到设计感很强的仓库所在的夹层，从会议室观察，整个公司尽收眼底，楼下是绿色和蓝色的四边形工作台，公共会议区的下面是带黄黑警示条纹的总监室。沿着夹层可以通过用钢和木建造的桥，经会议室到达后门。

"

设计师在办公区设计了一个双层仓库空间，如同在陆地上建一个港口。在这个仓库上插入了两座桥一样的悬空结构，同公共办公区相连。新办公室可以不设门，工作台的设计更加实用。穿过红色的展示墙就是主要办公区。公共会议室的墙面采用钢结构，董事会会议厅由几面由悬挂的冲浪板组成的 2 米高的墙隔开，旁边的休息区设有一扇 1.6 米高的升降门同户外庭院相连。

Our strategy of inserting double height warehouse-like space in the office creates a harbor in the land. Two floating structures are inserted in the warehouse to accommodate rooms needed to support the adjacent open work areas.

Google 总部办公室

设计公司：Clive Wilkinson 建筑公司
项目地点：美国加州
项目面积：63240 ㎡
摄影师：Benny Chan/Fotoworks

项目介绍 PROJECT DESCRIPTION

Google 的创始人 Larry Page 和 Sergey Brin 在斯坦福大学的学生宿舍内共同开发了全新的在线搜索引擎，然后迅速传播给全球的信息搜索者。Google 目前被公认为是全球规模最大的搜索引擎，它提供了简单易用的免费服务，用户可以在短时间得到相关的搜索结果。

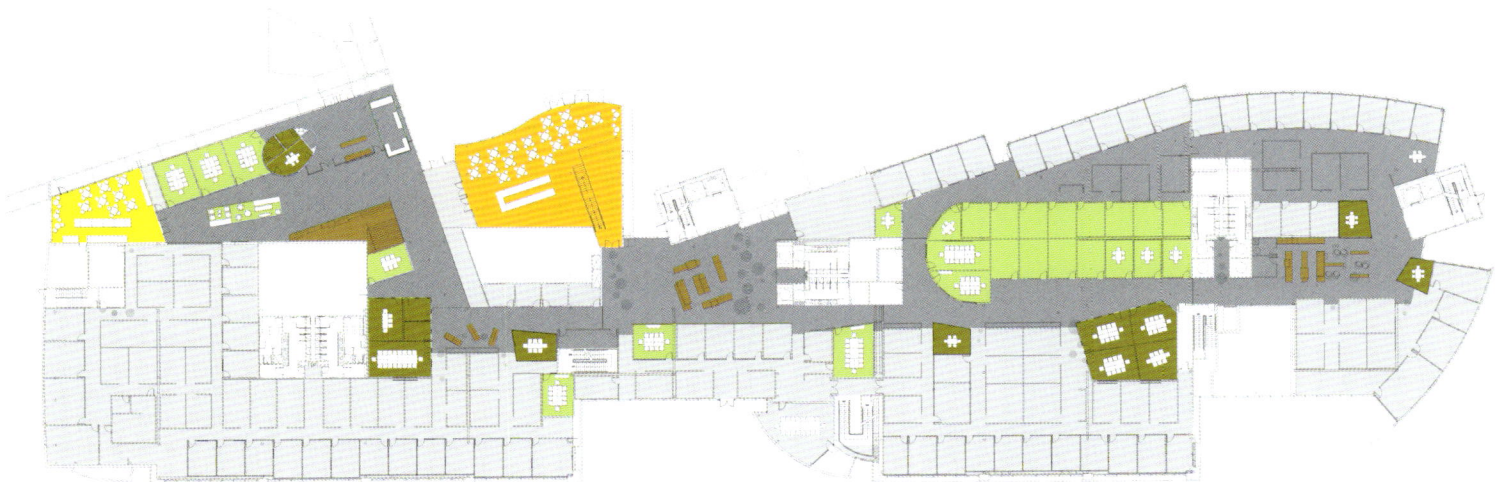

早在 2004 年，Google 公司选择 CWA 设计公司一起参加设计竞赛，为其在美国硅谷的 46 451.52m² 软件工程园制定了宏伟的设计目标，他们与 DEGW 国际设计顾问公司和著名生态建筑师 William McDonough 共同做出总体规划图，旨在构建一个多样化的园区环境。

In early 2004 Google selected CWA in an invited the design competition, setting ambitious goals for the design of its 500,000 SF Silicon Valley campus. Together with workplace strategists DEGW, and cradle-to-cradle environmental visionary William McDonough, a master plan was developed to create a diversified campus environment.

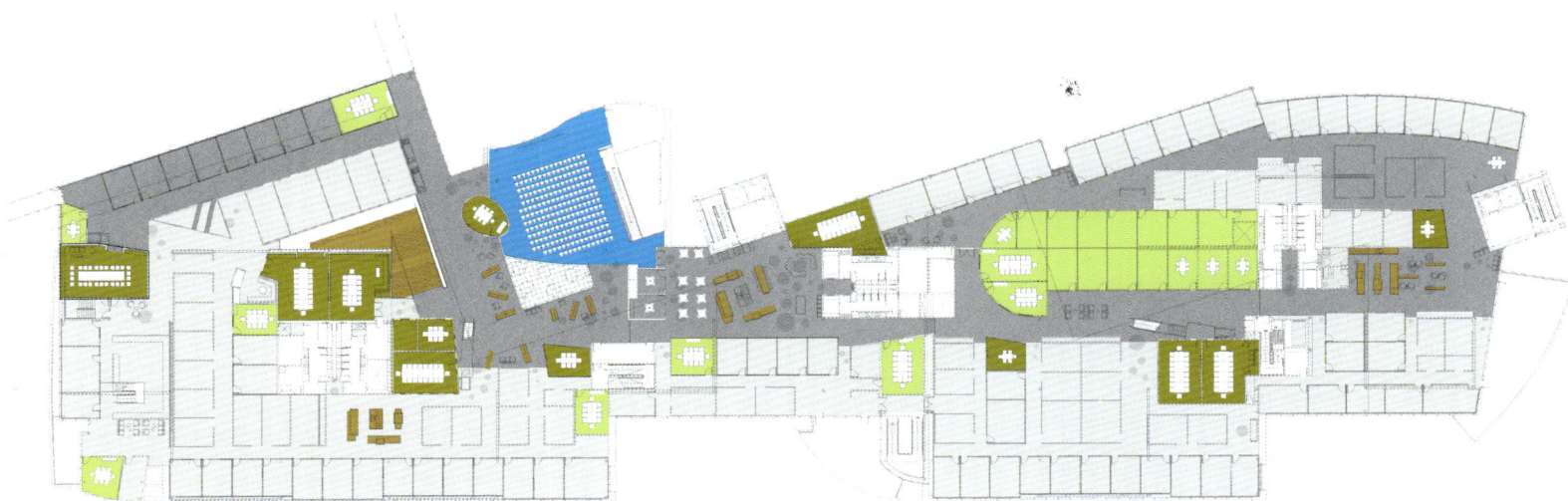

Second Floor Plan　二层平面图

设计考虑了现有的内部结构和外部建筑结构，从整体上构建了一个集学习、会议、娱乐、餐饮为一体的软件工程工作环境。

The design integrated highly focused software engineering workspace with learning, meeting, recreation and dinning facilities, taking into consideration the existing conditions of the inner courtyards and building shell.

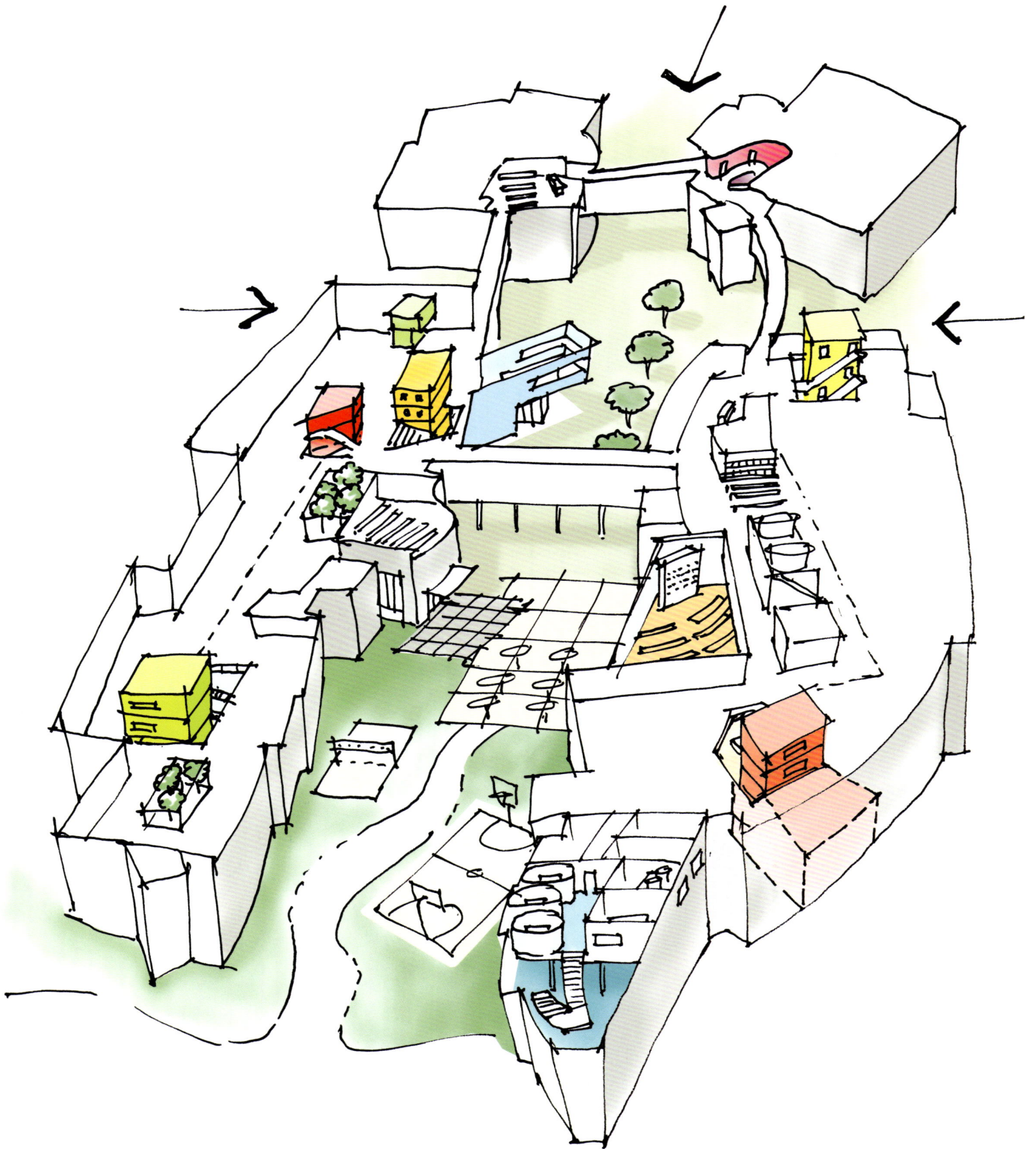

总规划图根据工作"社区"和"主街道"的位置进行简单设计。所有的公共空间都分布在"街道"两旁：会议室、技术研讨区、微型厨房、图书馆和娱乐室。在许多三人玻璃工作间中，天花板由吸音材料制成，隔音的同时也将光线反射到室内深处。提供餐饮服务的咖啡馆和分布广泛的微型厨房是"社区"的社交中心。这里还有休息室、图书室以及装修过的会议区和集会点。公共道路沿线的特定区域被设计成技术讨论区，人们经常聚集在此进行技术研讨、知识共享。

环保顾问的参与使可持续发展观念得到重视。项目运用了大量可回收产品和材料。该建筑物的原配门被用作装饰物，所有的开放式工作台都是可回收的。

The Masterplan followed a simple distribution of work "neighborhoods" along a "Main Street" circulation plan. All shared resources were located along this street, and ranged from meeting rooms, to tech seminars, micro-kitchens and library lounges. Several different food service cafes were provided and distributed, and micro-kitchens were opened up to function as social hubs with lounge facilities, library areas and furnished open spaces to encourage spontaneous meetings and gatherings.

Following the participation of the environmental consultant, a sustainable energy-conserving environment was a high priority, and most building materials used either were cradle-to-cradle products, or contained high-recycled content.

MTV NETWORKS

MTV 音乐电视网德国柏林总部

设计公司：Dan Pearlman 设计公司
项目地点：德国柏林
项目面积：280 ㎡
摄影师：diephotodesigner.de

项目介绍 PROJECT DESCRIPTION

MTV 音乐电视网是世界音乐娱乐网络巨头，在全球拥有 160 余个网络频道，其业务领域包括电视、数字媒体、出版、家庭视频、广播、唱片、娱乐、版权营销等。公司在全球积极地与超过 500 个互联网宽带、无线通讯、互动电视公司合作，扩大其影响力与业务范畴。

设计师运用设计的魔力，创造了一个"灵感的中心"。将整个空间南北朝向的中心区域扩展开来，把宽阔的中庭变成了一个"品牌标志展厅"，其中配备了公共座椅，可供 250 名员工在其中举行正式会议。

The designers of Dan Pearlman were asked to perform their magic and created an "axis of inspiration" by opening up the entire centre area on the first floor from the north to the south façade. Today the spacious atrium serves as a "Brand Garden" with seatings for taking breaks or informal meetings, which can accommodate up to 250 employees.

该公司的设计风格和用材都与 MTV 德国总部的 LOGO 相统一，极富特色，并运用了褐、白、黄色调。前台、休息室、咖啡厅、中庭和小厨房都设计得很有创意。

The interior design and materials relate to the identity of the MTV Networks (Germany) logo with its characteristic font and the colours dark brown, white, and yellow. They are referenced and translated creatively in the design of the reception area, the lobby, and the café as well as in the atrium and the kitchenettes.

设计师利用空间的自由感和各种各样风格独特的私密区域营造出一个具有生活感的工作室。员工可以选择在员工餐厅、蓝色休息区、标志展厅的装饰篷下就餐或小憩。运动休闲室还设有乒乓球台和桌面足球。树状阳伞、几处座位与入口处的装饰材料点缀着空间。宽阔的中庭设计有公共设施、直立的绿色植物、如由景观石建造的座位等一些装饰细节。

Greater freedom of movement and an equal variety of unique as well as private retreat zones contribute to a better work-life balance at the workplace. Whether having lunch in the Network Kitchen, in the Blue Lounge or in the Brand Garden under a stylized canopy of leaves, there are ample choices for taking time out during the workday. Fans of table tennis or table football can of course be found in the Sport Lounge. The abstract tree parasols and several different seating isles refer to the self-contained "corporate" design and use of materials in the entrance area. The overall picture of the "garden" is rounded up by the interaction of the large airspace - leaves installation and the "vertical green" spaces as well as small details, such as landscape stones as open seating isles.

美奎地产公司办公室

设计公司：Clive Wilkinson 建筑事务所
项目地点：美国
项目面积：1339.2 m²
摄影师：Benny Chan/Fotoworks

项目介绍 PROJECT DESCRIPTION

美奎地产公司是一家提供全面服务的房地产开发公司。其运营策略是针对位于美国东部的办公楼、工业资产的收购和开发机会做出迅速反应。凭借该公司对多种资产类型和市场动向的把握，从而让合作方和投资者获益。

受加州南部海滩本土文化和周围环境的影响，设计语言采用了两种波浪的形状来表现水的流动。行政会议室位于办公区的两端，倾斜的墙面由碎片状墙板构成，中面夹着一层深浅不一的彩色玻璃，形成质感的多面体，如同飞溅的浪花。开放的工作区中重叠放着许多圆形的隔板，让人联想到此起彼伏的海浪。

Inspired by Southern California's native beach culture and the site's natural surroundings, the main design language employs two different variations of curved shapes, expressing the extremes of the motion of water. Anchored at both ends of the space, the executive conference rooms are made up of edgy, shard-like slanting walls. Colored glass in dark and light shades is sandwiched in-between, creating a highly textured and faceted surface that resembles breaking surf. Throughout the floor in open workstation areas, more rounded overlapping low partitions evoke imagery of gentler rolling waves.

两个会议室和工作区之间由流线型的天花板连接，上面装有照明，楼顶的管道也隐藏于此。运动和能量元素也在房间的每个细节中体现。前台、茶水间、会议室运用木制品作为主要装饰，使整个设计风格统一。

该设计使用了数控技术制作 1:1 的泡沫模型，对造型设计进行比对调整，保证了复杂曲线设计的精确度。这不仅有利于测算，也避免了材料的不必要浪费。

The two conference rooms and workstations in between are connected by a floating ceiling plane. The ceiling is a dynamic visual element that houses lighting attach and also visually conceals the mechanical ducting located above, while allowing the overall building envelope to be exposed beyond. The expression of motion and energy is also carried through in the detailing of the space.

Full size CNC-milled foam mock-ups are used as a design and construction tool on site to fine tune the highly sculptural design and ensure precise construction of the multiple unique complex curving forms. The foam pieces not only contribute to time and cost savings in construction, but also serve as a life-sized model for the architect.

KAO CORPORATION

花王株式会社办公室

设计师：神山和裕、堀川纯一
设计公司：HaKo Design
项目地点：日本东京
项目面积：1537.5 ㎡

项目介绍 PROJECT DESCRIPTION

花王株式会社是日本一家生产日用品的公司，总部位于东京，主要的业务是研发、制造与销售清洁用品。虽然近年来领域趋于多样化，但花王仍然以油脂化学、界面化学和高分子化学作为其发展的三大支柱产业。其创立目标是为满足人们对美、洁和健康的需要，一直致力于创新技术研发并将最先进的技术运用于生态项目。

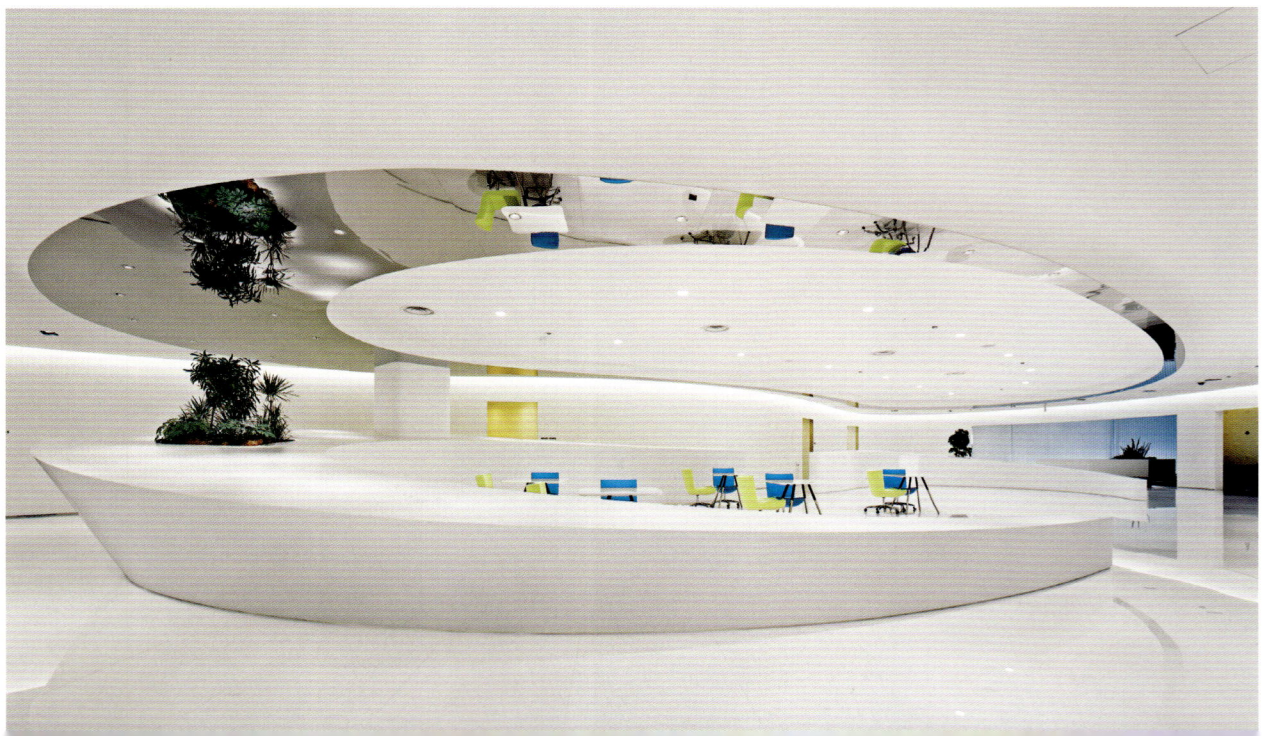

"

设计师根据办公室形象和发展方向营造出一个柔和、淡雅、流畅的办公环境。整体是白色、精美、柔和的曲面体，如同人的皮肤一样温暖而柔软，应该将它定义为一种有机设计。

Keeping in mind the company image and policy,the designer decided to design its offices in harmony with those features, therefore, we opted for an atmosphere that was soft, mild and smooth. So, the leitmotifs here were white, beautiful, seamless and gently curved surfaces, evoking the same warmth and tenderness of one's body skin. We'd like to think of it as a sort of an organic design.

穿过一条长而窄的走廊到达办公区，动态的分区位于这里的开阔区域，依据视角和透视原理给人带来不同的观感。这个区域由传统石膏工匠做成日本传统石膏墙面，平滑而洁白，同时表达了公司对健康的关注。除此之外，等候室和其余区域由建造的剩余材料建成，营造出森林的气氛，体现公司对环保的重视。

Especially, the dynamic partitions that are placed in the area which widens after passing through the long and narrow corridor leading to the office space, are created in such a way that they provide different impressions depending on the viewing position and viewers' perspective.

每个会议室的布局都不同，这是为了在会议中激发员工的创造力，提高积极性，同时兼具原创性和功能性。会议室的天花板是阳光的图案，室内安有的玻璃板减少视觉的狭窄感。完全暴露的混凝土材料与铺设的地毯在视觉上形成强烈的对比。白板构成一条轴线，两间房间依轴线设计，相互映衬。

In order to avoid a feeling of narrowness, the conference room is arranged with sunbeams graphic on the ceiling and a glass panel, this entire room has been thought to suggest the ambience of a garden. The main feature is the contrast between the fully exposed concrete material and the warmth of the fitted carpet. Also, the room where the white board becomes an axis and two divided rooms by this axis are designed to be a reflection of another half.

CISCO SYSTEM

思科公司驻意大利办公室

设计公司：Progetto CMR Massimo Roj Architects
项目地点：意大利
项目面积：5000 ㎡
摄影师：Benny Chan/Fotoworks

项目介绍 PROJECT DESCRIPTION

思科系统公司 (Cisco System, Inc.) 是全球领先的互联网设备供应商，也是美国最成功的公司之一。1986 年生产的第一台路由器，使不同类型的网络可以可靠地互相连接，掀起了一场通信革命。思科公司每年投入 40 多亿美元进行技术研发。过去 20 多年，思科几乎成为"互联网、网络应用、生产力"的同义词，思科公司在他们进入的每一个领域都成为市场的领导者。

这个 5000 平方米的空间设计的关键词是透明、光照、色彩。

布局的设计灵感源于电脑主板，上面插满了可以移动的零件，也包括"数据"传输交流原件。同时，多元化的办公室环境中包括许多开放和关闭式空间、交互式的走廊、休息室以及一些非正式集会点。

企业形象是通过从前台贯穿整幢大楼的一系列符号表现的："交流、科技、互动、人力网络"。每一层楼都有不同主色，在促进交流与协作中起到不同的作用。

" The guiding principles in the interior design of the 5000 square meter building are transparency, light, and color.

The floor layout is inspired to the design of a computer's motherboard. Just as a motherboard includes fixed and movable parts, as well as elements for "data" exchange and communication, in the same way, diversified office environments are created, including open spaces and closed spaces, connected by interactive corridors, relax areas, and areas for informal meetings.

The corporate image is conveyed in the reception and everywhere in the building through a series of symbols representing Cisco's founding principles: communication, technology, interaction, and the human network. Each floor in the building is characterized by a different color and accommodates a different company function, to facilitate communication and work flows.

Changing the Way We Work, Live, Play, and Learn.

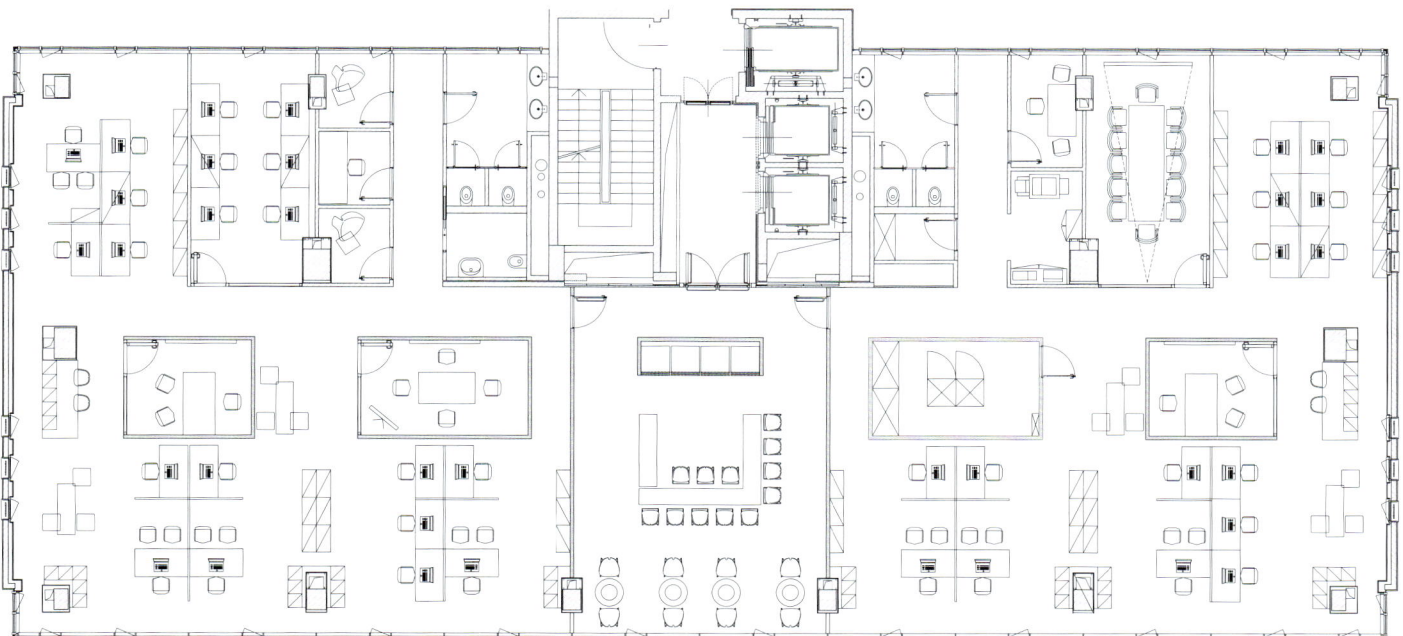

"

底层用于会见客户和来访者。咖啡区中摆放着的纯白家具与颜色艳丽的椅子和创意灯形成强烈的对比。这层楼还配置了多功能房间，用于会议、研讨、培训，还可以根据需要隔成两个房间。

The ground floor is dedicated to receiving guests and clients, with a coffee area characterized by the use of pure white furniture and vividly colored chairs as well as innovatively designed lamps. On this floor there is also a multifunctional room designed to accommodate presentations, workshops, and training courses, which can be divided in two smaller rooms, according to needs.

the human network effect

服务器机房及远程技术汇报中心位于一楼，这里用于展示思科加利福尼亚总部研发的新产品和新系统。思科的创新通信技术应用于三个多媒体办公室和一个视频会议室中，视频会议室应用的是思科自主研发的创新系统。

The first floor accommodates the server room as well as the Remote Technical Briefing Center, an area to experience new products and systems developed at Cisco's headquarters in California, thanks to the company's innovative communication technologies, and three medium conference rooms, as well as a video conference room with an innovative system patented by Cisco themselves.

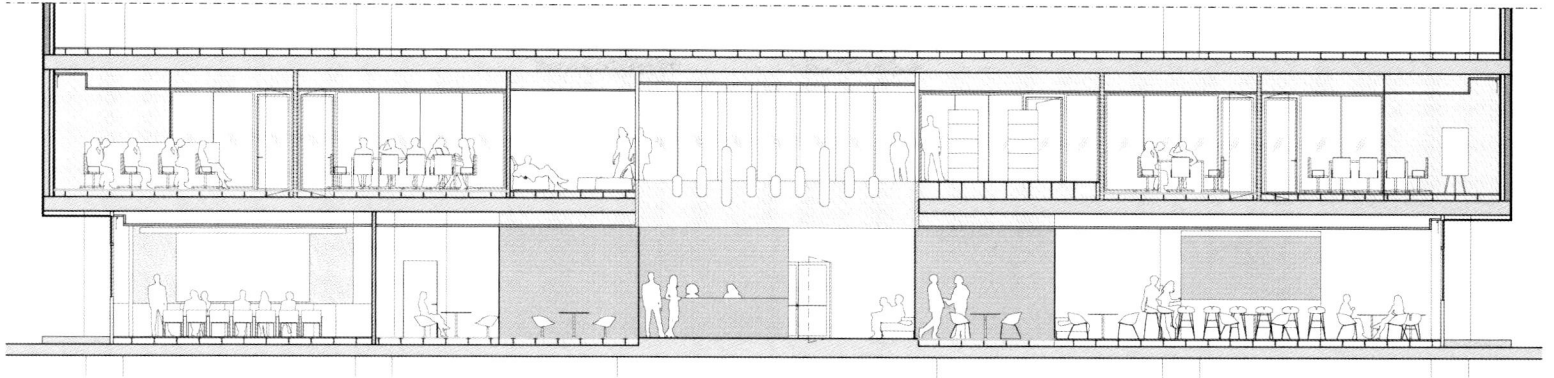

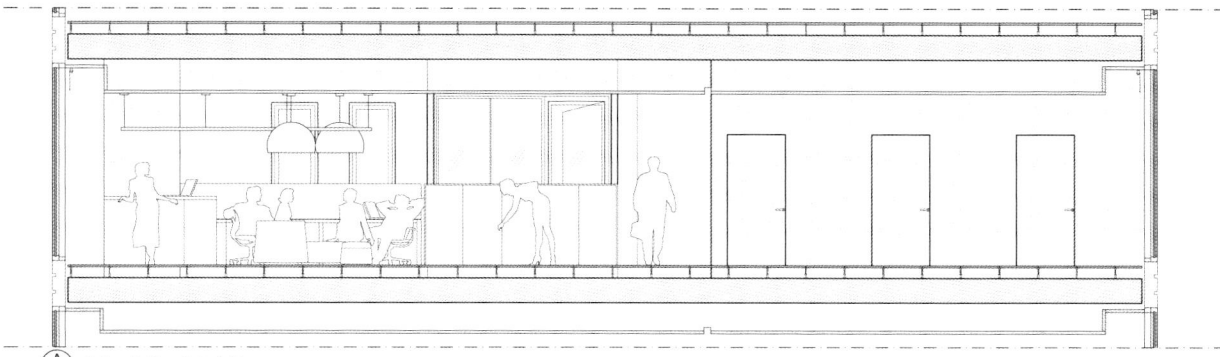

A
EA101 Sezione A-A' - Scala 1:50

A
EA101 Sezione A'-A - Scala 1:50

Sezione B-B' - Scala 1:50

Sezione B'-B - Scala 1:50

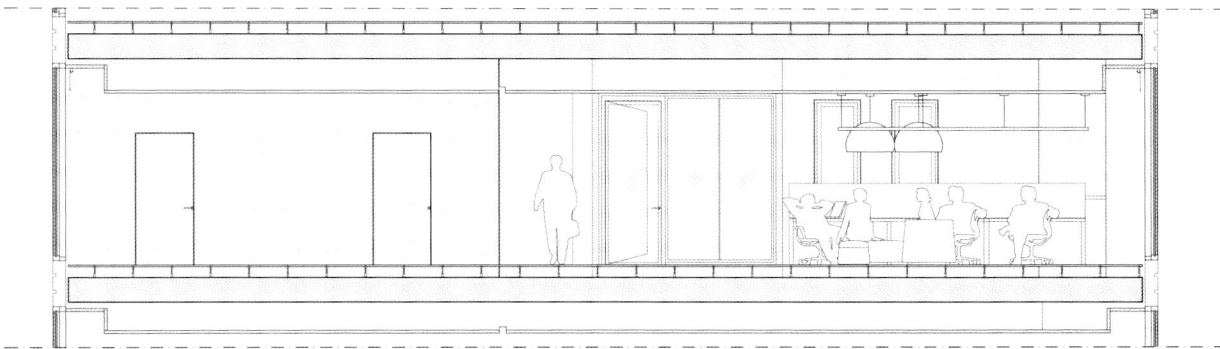

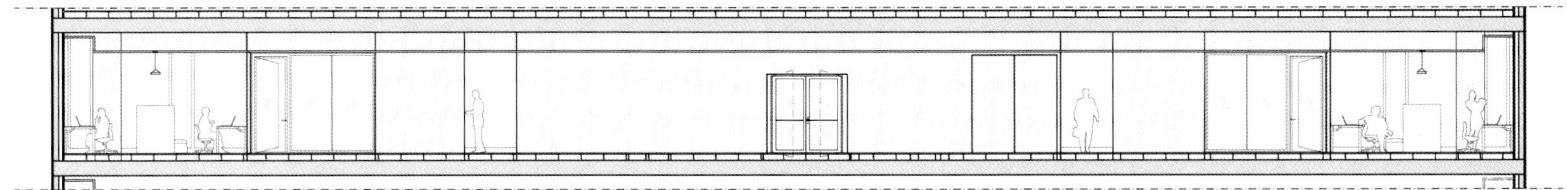

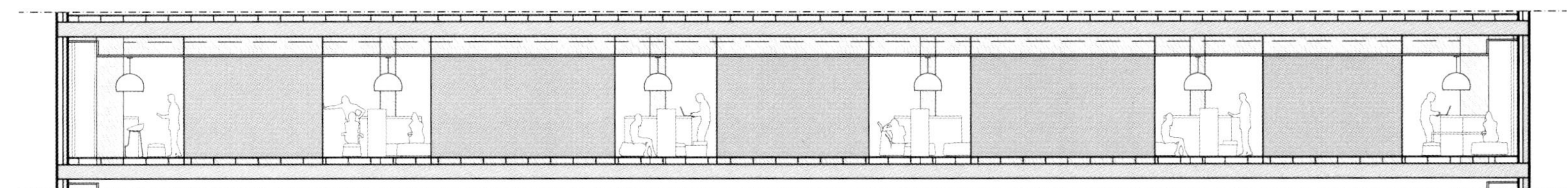

THESE DAYS

These Days 广告公司

设计公司：Creneau International
项目地点：比利时
摄影师：Philippe Van Gelooven

项目介绍 PROJECT DESCRIPTION

These Days 隶属于伟门广告公司，通过网络拓宽市场，专注于品牌互动，其目标是积极寻求合作伙伴并扩大在全球广告界的影响力。目前该公司在荷兰的安特卫普和阿姆斯特丹设有分公司，吸纳了多元文化背景的广告精英，客户遍布世界各地，推广产品从手机到香蕉无所不包。

These Days 位于比利时的一个旧汽车工厂。Creneau 国际为 These Days 设计了一个现代装饰艺术风格的办公空间。

These Days, a communication firm specialized in "brand interaction" through internet marketing and advertisement, is situated in an old car factory in Antwerp. Creneau International created an atmosphere where great urban influences were combined with art-deco style.

入口处一个大显示屏用图片营造氛围。进入大厅，就像进入俱乐部，各种材料和缤纷的色彩将你带入一个神秘的地下空间。超大的会议室放有两个玻璃做成的黑白方块，营造了独特的氛围。

The entrance, where a huge screen introduces the image culture, is understood as the club lobby and made of materials and colours which create an underground atmosphere. The monumental meeting rooms, two black and white cubes made out of glass, create an alienated environment.

另一个亮点是圆柱形的非正式会议室布置得像古代庙宇，3 米高的墙上覆盖着羊毛和皮革。会议室用来激发头脑风暴。黑色调的走廊中布置着黑色的灯。树木装饰让整个空间充满自然的气息，让人感觉精神饱满。

The other eye catcher is the informal cylindrical meeting room which resembles the ancient temples. The walls (3 m height) are covered with wool or leather and this meeting room is an inspiring setting for brainstorm sessions. The black hallway is enlightened with black lights. Trees make sure that the space is filled with nature and freshness.

MOTHER

Mother 广告代理公司

设计公司：Clive Wilkinson 建筑事务所
项目地点：英国伦敦
项目面积：1302 ㎡
摄影师：Adrian Wilson

项目介绍 PROJECT DESCRIPTION

Mother 广告代理公司在伦敦、纽约和布宜诺斯艾利斯均有分支。该公司是伦敦最大的独立广告公司，是世界公认的最具创意和活力的广告公司之一。公司信条是：工作第一，乐趣第二，利润第三。

从 6 个人的小公司发展成为目前英国首屈一指的广告公司，Mother 广告代理的秘诀就是将业务与现代文化融入到他们的工作环境中，形成一个平面结构的组织，没有空间上的等级特权，每个人都围着一张桌子就座，随着这张桌子的扩大，公司也在渐渐成长。

In six years, Mother has grown from a six-person boutique ad agency to Britain's leading agency. Their radical approach to the advertising business and contemporary culture is translated into their work environment. A flat organization with no privileges, everyone works around a single large worktable. The advertising agency will grow the table as the company grows.

为了让大厅和下面的两层楼更紧密地相连，设计师在横贯大楼的小路上设计了新的混凝土楼梯，以连接三层楼。4.3 米宽的台阶延伸为混凝土工作台，像跑道一样环绕在第三层楼，这可能是世界上最大的桌子，76.2 米长，最多能容纳 200 个人。混凝土工作台的设计灵感来自于都灵菲亚特汽车厂房屋顶跑道，是广告界的一个里程碑。

In order to achieve a strong connection to the Loading Bay Lobby, two floors below, the architect proposed to build a new concrete staircase on the width of a small road cutting through the building to connect the three floors. This 4.3m wide staircase would turn into the Agency's cast-in-place concrete worktable and circuit the 3rd floor room like a racetrack, becoming perhaps the world's largest table at 76.2m long, broken in sections for circulation, with a maximum capacity of 200 people. The inspiration for the concrete table was the iconic 1920's Giacomo Matte-Trucco roof top race track for Fiat Lingotto in Turin, making a monumental statement about the most ephemeral of arts: advertising.

整个工作室以中性色调为主，白色树脂地板。为减少厂房的噪音，CWA 设计公司用 75 毫米厚的吸音泡沫材料制作了 2.1 米长的灯罩。50 个不同形状的灯具用工厂收藏的 50 种不同花纹的布料做成，有很好的艺术效果。在其他楼层上，专门定制的塑料挂帘隔出不同空间，供其他部门和子公司使用。

On all floors, all surfaces were painted white, with a white epoxy floor, to achieve a neutral art studio space. To mitigate sound in the hard factory space, CWA designed 2.1m long lamp shades padded with 75 mm of acoustic foam. 50 light fixtures were then covered with unique patterns of Marimekko fabric selected from archive stock in its factory in Helsinki, achieving the effect of a large art installation.

ENGINE OFFICES

Engine 国际营销公司伦敦办公室

设计公司：Jump Studios
项目地点：英国伦敦

项目介绍 PROJECT DESCRIPTION

Engine 是一家国际整合营销传播公司，提供广泛的市场营销传媒解决方案。公司已从伦敦到纽约发展到许多国家。客户包括宝马、葛兰素史克、MINI、诺基亚西门子网络、劳斯莱斯、索尼、沃尔图等。

受公司老总的启发，项目采用了"精密工程"的概念："我们希望办公室看起来像是机械制造出来的，而不是人修建的。"从一楼到五楼，机械感无处不在。最突出的是入口处流线型的礼堂。"从造型的角度来说，非常好。这当然也是客户的视线焦点。"设计师说。

The team worked with the concept of "precision engineering", partly inspired by the chairman of Engine company, Peter Scott, "We really wanted the office to look like it was machined rather than constructed."

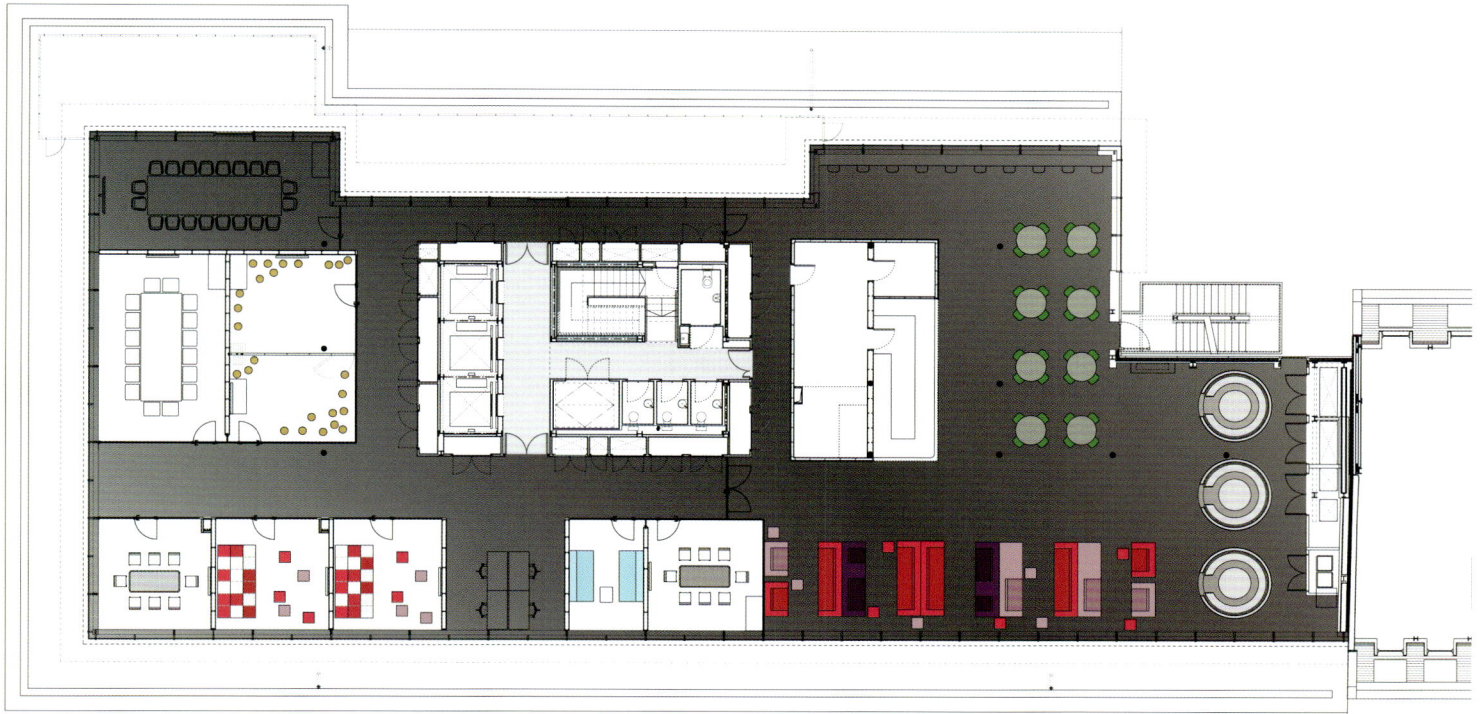

作为一个整合了 10 个不同行业的 12 个领先公司的国际营销公司，Engine 的要求无疑是苛刻的。最大的挑战就是文化差异的统一，从年轻的广告创意公司到政党宣传公司，设计师必须克服这种差异，让每个公司都认同。

With 12 different companies operating under the Engine umbrella, the challenge is to create an environment that would appeal to a broad range of tastes while respecting and upholding the individual brand identities. "We had to cater for a spectrum of cultures," says Jump director. "That ranged from typical creative young ad agency through to political lobbyists who would be horrified to think that someone might come to work in a pair of trainers."

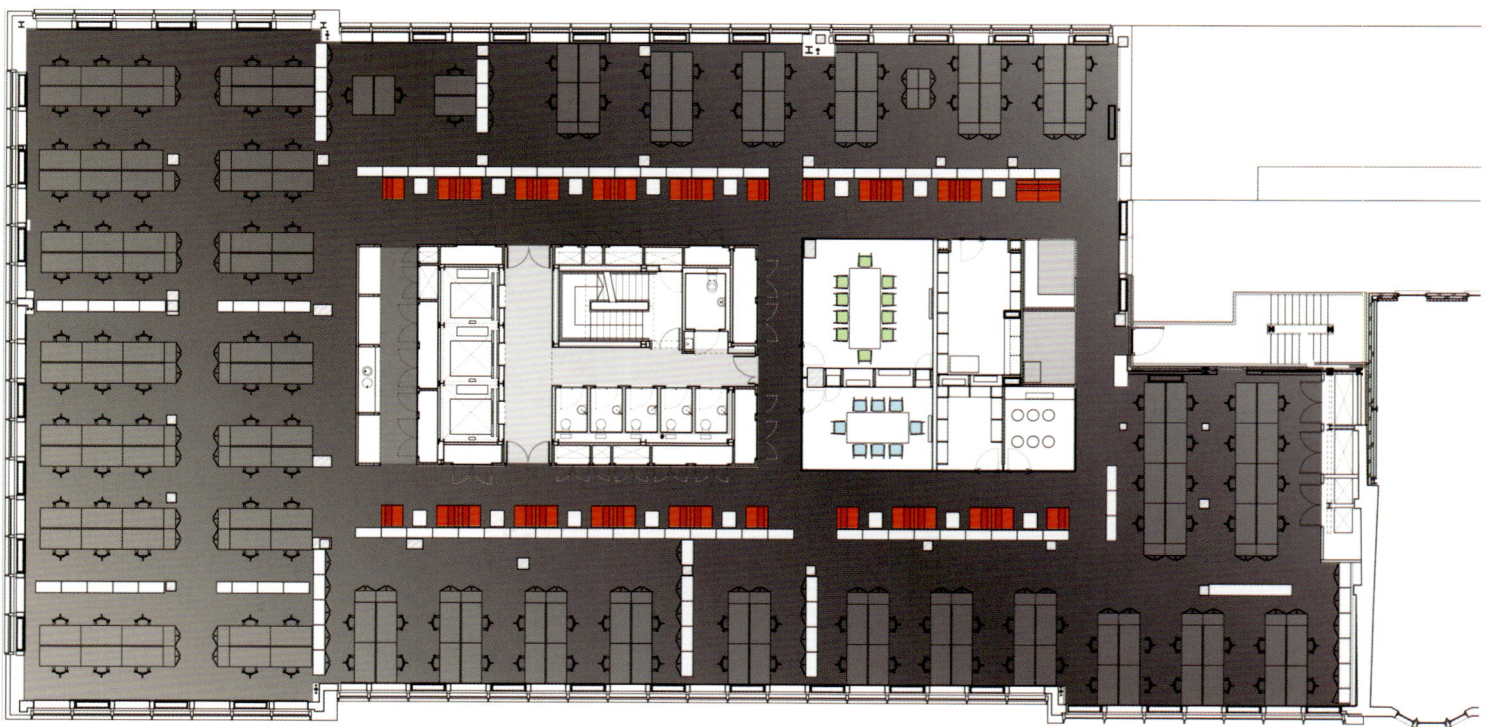

项目其他可圈可点的地方也比比皆是。五楼的创意座椅搭配可丽耐外壳和巴力天花板。有了可以俯瞰城市的顶楼咖啡座、新潮而宽阔的会议室等设施，员工们更愿意相互交流了。这就一种是简洁而永不过时的设计。

Among many other "talking points" in the building are the seating pods on the fifth floor with their Corian shells and Barrisol light ceilings. Here employees are encouraged to interact, serviced by a café offering spectacular views across the city's rooftops and a series of conference and meeting rooms ranging in design, size and style.

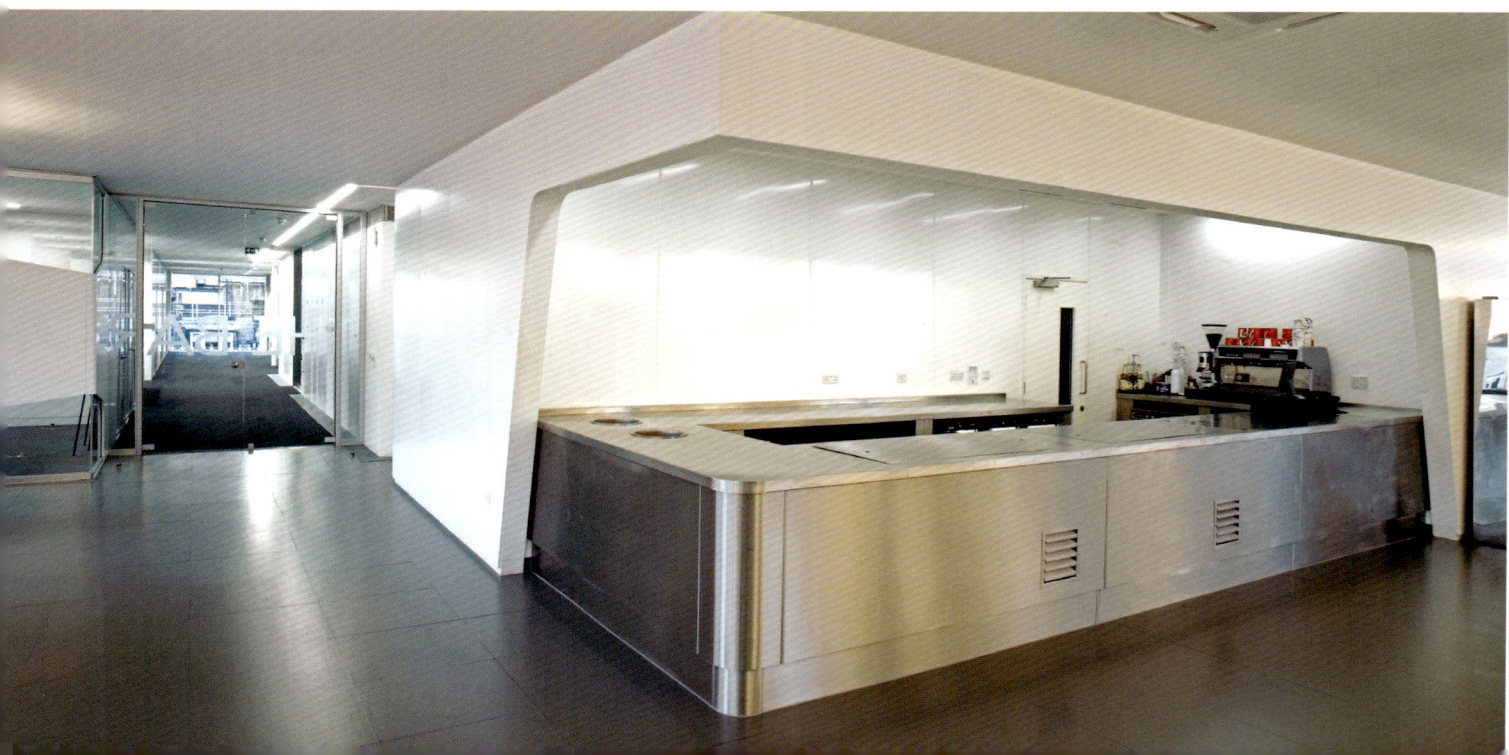

NIJE GRITENIJE

荷兰 Nije Gritenije 基金会办公室

设计公司：FLATarchitects
项目地点：荷兰
项目面积：370 ㎡
摄影师：Arend Loerts

项目介绍　PROJECT DESCRIPTION

荷兰 Nije Gritenije 基金会的发起方是荷兰拉博银行，又称农业合作银行，是在 1973 年由荷兰数家农村信用社合并而成的。该银行是荷兰的第二大银行，在世界各大银行中居第 31 位，自 2001 年起被全球金融杂志评为全球十大最安全的银行之一，在世界 500 强中排名第 175 位。该企业坚信开放透明的文化是其与合作成员建立诚信、实施管理和监督的保证。

该基金会是荷兰拉博银行为了鼓励当地区域的创业精神成立的。

走上九楼的梯子，迎面而来的是四个不规则形状的造型摆设，将入口分隔开，通过上面的空隙，也可以看到后面的工作区。通过缝隙和光面墙，来访者的视线很容易被木质装饰吸引，装饰物的另一面则是书架。

入口旁边有一个固定工作台。四张工作台从一个组合柜延伸出去，颜色与可移动的工作台不同。四张大桌子为 16 个临时员工提供办公地点。桌子由一座钢质框架和一张钢质网格台面构成。根据不同的用途，这张特殊的桌子配备了不同大小的桌面、复式电插孔、储藏台等功能元素。

The stairs towards the ninth floor end in an entrance space fenced off by four irregularly shaped objects. Together they form a "porous wall" filtering the activity in the work space behind. Through the gaps and reflecting surfaces on the wall, sounds and images are presented to the spectator in fragments.

The permanent workstations are positioned next to the entrance. The four desktops fan out from a shared drawer unit. Their shapes and colours differ from the flexible workstations. Four large tables together provide space for sixteen temporary workers. The tables consist of a steel frame with a tabletop made of steel grid panels. The tabletop can be arranged according to ones' needs by combining various desk elements, such as desktops of several sizes, storage objects, book displays and multiple sockets.

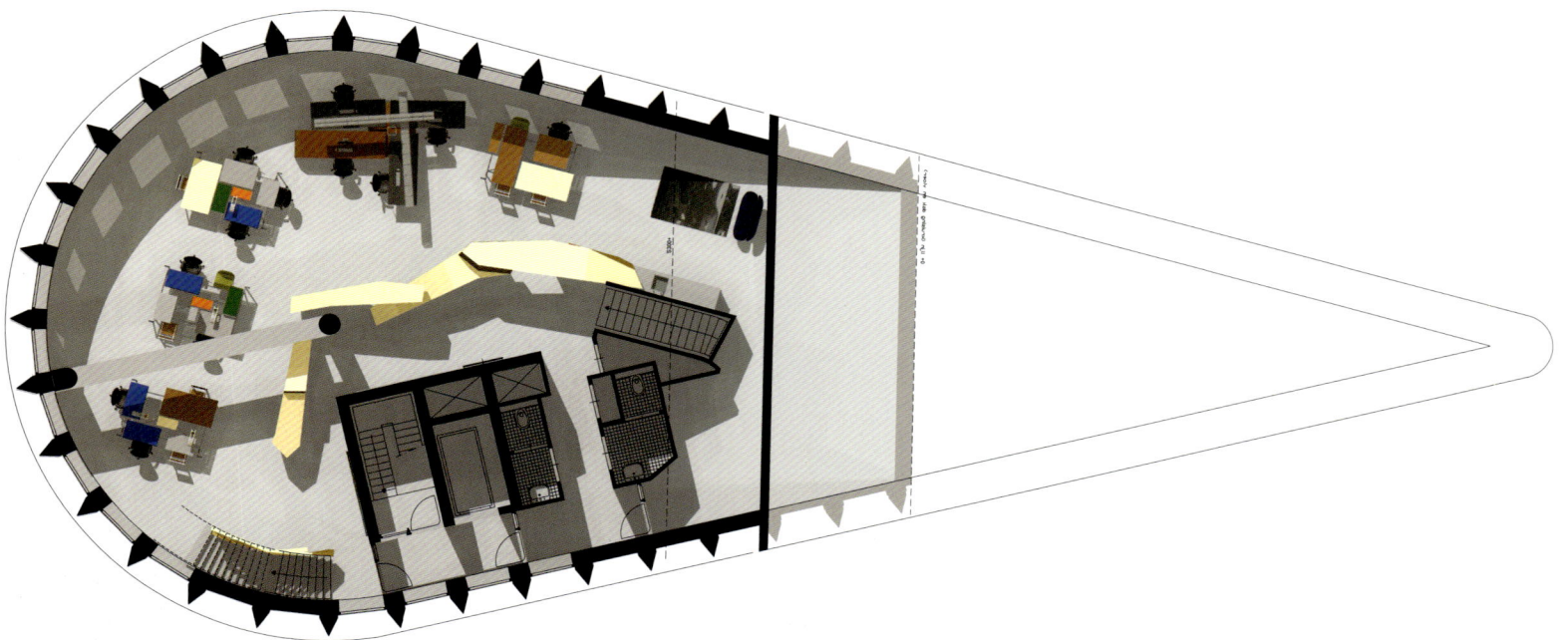

十楼是会议室和演示厅。来到这里，你会很容易被地毯和被它盖住一部分的墙面上的图案吸引——这是一幅巨大的荷兰弗里斯兰地图。基金会成员通常会聚集到地图上的会议桌，桌旁的 11 张椅子代表 11 座城市。拉上铺满墙面的窗帘，该空间可以用来放电影或者幻灯片，然而，当你往窗外眺望时，窗户外的景观则是一幅真正的弗里斯兰城市地图。

From the entrance space stairs lead on to the conference, and presentation room on the tenth floor. Upstairs your attention is immediately drawn to the carpet, which is also covering parts of the wall. It shows a giant map of Friesland (the Dutch province) designed by graphic artist Martin Draax. The members of the Nije Gritenije foundation gather over this map, around a grand conference table with eleven chairs, one for each Frisian city. By closing a wall-to-wall curtain, part of the space can be dimmed for showing films or presentations. When looking out of the windows, the real map of Friesland unfolds.

RED BULL

红牛品牌总部

设计师：Shaun Fernandes、Simon Jordan、Go Sugimoto
设计公司：Jump Studios
项目地点：英国伦敦
项目面积：1860 ㎡
摄影师：Gareth Gardner

项目介绍 PROJECT DESCRIPTION

红牛（Red Bull）是全球著名能量饮料品牌。老板马特希茨长期在品牌推广上进行经营，通过赞助极限运动、汽车运动和飞行运动等建立了品牌声誉和影响。在各国的推广也颇为成功，目前已经行销 140 多个国家和地区。

红牛新办事处占据了一座 19 世纪建筑物顶楼的三层，屋顶处的外阳台在"玻璃盒"形状的中心建筑上延伸，伦敦西区在这里尽收眼底。

The brief is to amalgamate two separate offices into one central headquarters building. The new offices occupy the top three floors of an existing 19th century building, including a recent roof-level extension which takes the form of a glass "box" surrounded by an exterior terrace, providing spectacular views of the West End.

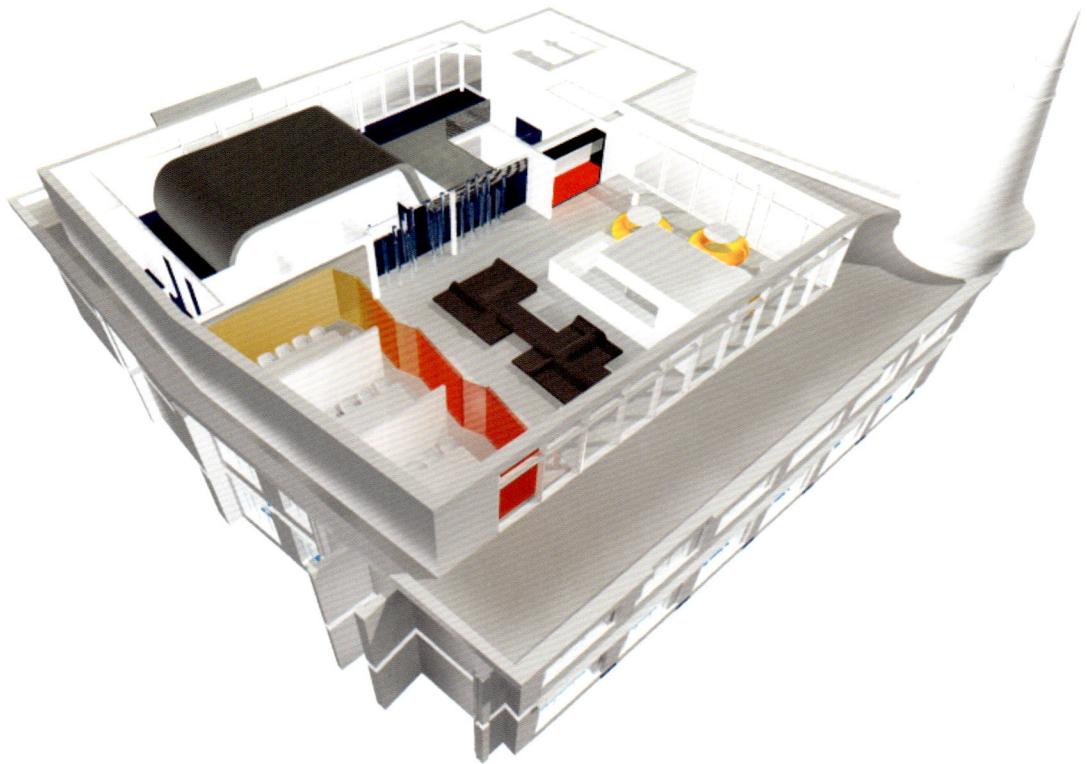

建筑的中空结构加强了下降的感觉，让人们眩目其中。中空部分的其中一面由一个三层楼高的视听墙占据，而另一面则通过一个悬浮的楼梯和一个滑梯形成戏剧化的循环，使得人们可以在各空间之间自由行走。

顶层的地板同延绵弯曲的碳纤维元素"缝合"在一起。这个元素在建筑的侧翼形成天棚，从外阳台延伸至整个建筑，包裹主会议室，划分出前台接待区，最后止于中空部分，形成了滑梯的围栏和楼梯的支撑。最后在低一层的空间以一个平台的形式结束，在那里形成了一个非正式的会议区域。

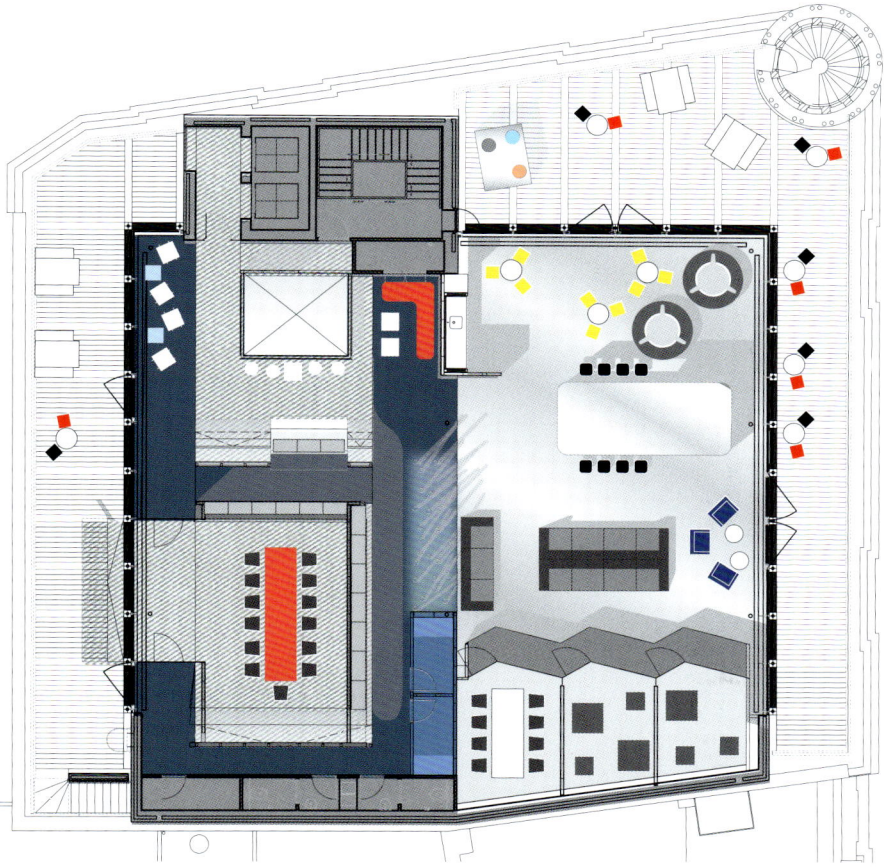

❝ This sense of descent is enhanced by voids punched through the building fabric, providing vertiginous views. Three-storey video wall occupies one void, while another includes dramatic means of circulation via a floating staircase and even a slide, aiming at encouraging free movement through the spaces.

These features of the top floor are "stitched" together by a continuous, snaking carbon-fibre element. It runs from exterior terrace (providing a wing canopy) through the building, encapsulating the main boardroom, forming the reception area, before disappearing through a void, forming the enclosure for the slide and support for the staircase. It terminates on the lower floor as a platform creating an informal meeting area.

LONG BARN STUDIO

长谷仓工作室

设计公司：Nicolas Tye 建筑师事务所
项目地点：英国贝德福德郡
项目面积：1000 ㎡

项目介绍 PROJECT DESCRIPTION

Nicolas Tye 建筑事务所的设计师们喜欢设计富有创意、健康环保的精品。从设计到施工，客户们对其工作都信任有加，因为他们在设计中总能找到新奇而简单的解决方案。该建筑公司遵循的原则是：不跟风，只设计清爽而健康的环境；无论何时何地，设计总能找到一个方法满足人们的需要。

長谷仓工作室位于英国贝德福德郡中部，莫尔登郊外。Nicolas Tye 建筑师事务所旨在为员工提供一个舒适、健康、富有激情的工作环境。该建筑一个重要的设计原则是让空间与谷仓本身以及与所处的环境相协调，与环境融为一体。工作室视野开阔，材料的运用简单而不失巧妙。

The Long Barn Studio is located within Mid-Bedfordshire, on the outskirts of the village of Maulden. Nicolas Tye Architects create a new-built studio space which could provide a comfortable, healthy and inspiring environment for up to a team of twelve. One of the key design philosophies is to harmonize and complement the existing structure of the barn and arable environment within where it sits. The studio itself is set down within the surrounding open landscape, and subtle materials which are used in a simple way reflect the local harmony.

"

建筑物整体是一个长方体结构。北面采用了无框玻璃,更多自然光的进入增加了室内光线,宽大的落地玻璃使工作室与周围环境融为一体。在工作室的南面分布着一系列像"豆荚"一样的、以松木板为材料的房间,避免了日晒温度过高的问题。每个"豆荚"都有专门的用途,如图书馆、复印室、卫生间、会议室等。科尔坦耐腐蚀钢的使用更突出了落地玻璃的设计和直线建筑结构的特点。室内采用石灰岩做地板材料,提高了室内轴线。工作室外墙全部由 20 厘米厚的松木板构成,内墙是砖块砌成的墙面。这样的隔热设计避免了室内温度变化太快的问题。

The studio is based around an elegant glass, rectilinear box. The "frame-less" glass panels allow high levels of natural daylight into the studio along the northern side and allow wide views of the surrounding landscape to become integrated into the studio environment. Along the southern elevation larch clad timber pods also punctuate the glass facades, which prevent from overheating. Each pod has a dedicated use including an architectural library, copy area, toilets and meeting room.

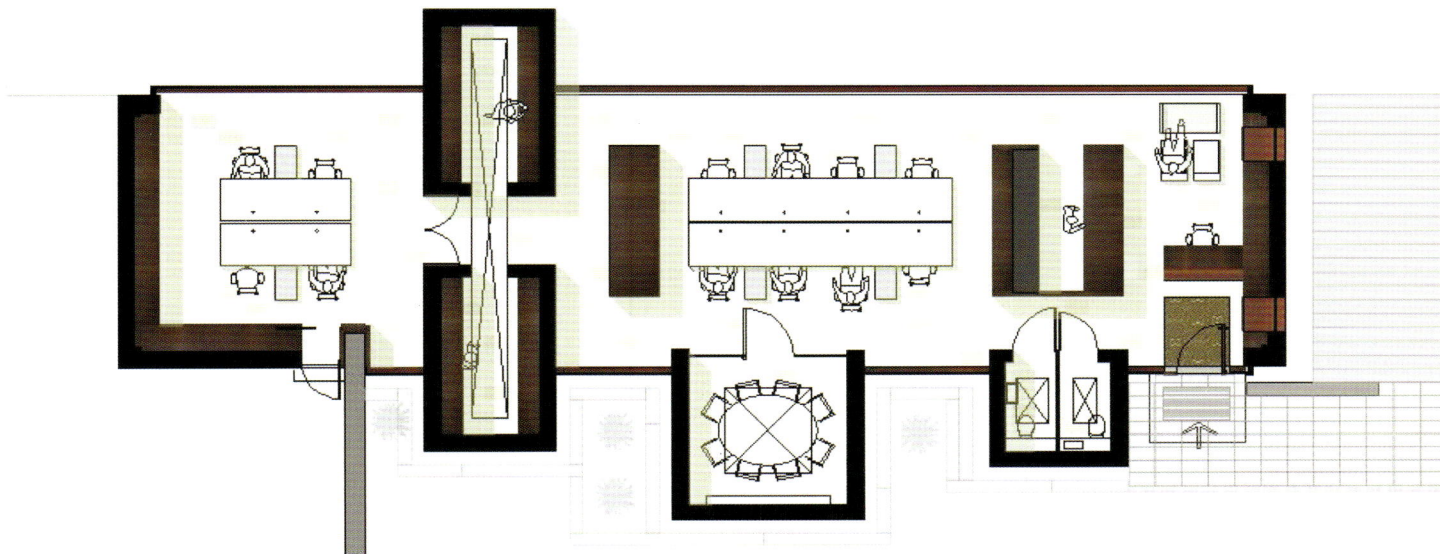

Plan　平面图

North Elevation　北立面

South Elevation　南立面

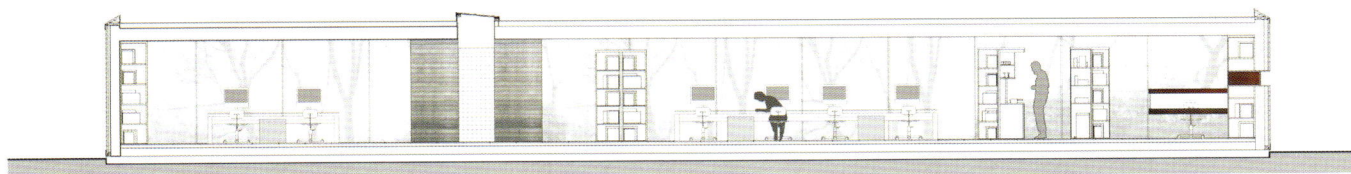

Section　剖面

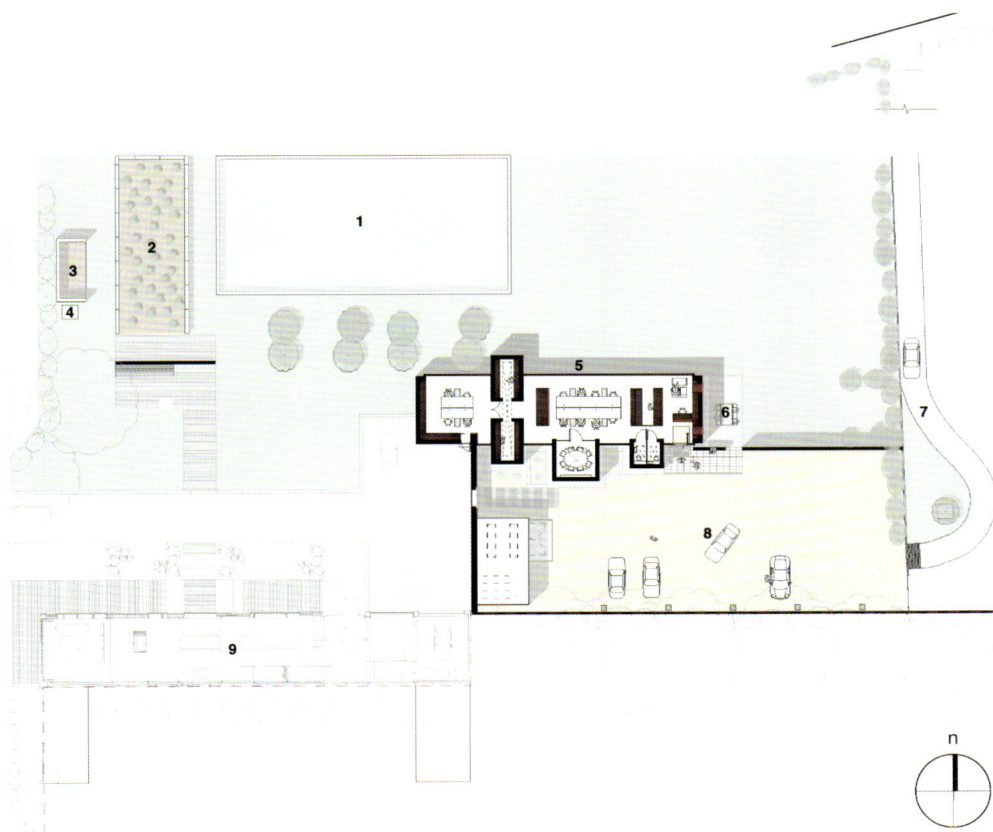

许多环保技术也贯穿于设计中，如风力发电机、雨水收集设备、低能照明控制系统、地暖系统以及空气热能交换循环系统。

Various sustainable and healthy technologies are also incorporated, including a wind turbine, rainwater harvesting, low energy centrally controlled lighting, underfloor heating system, a central vacuum and a combined heat exchange air source heat pump supplying natural fresh air.

ACCENTURE

埃森哲咨询公司意大利办公室

设计公司：Progetto CMR Massimo Roj Architects
项目地点：意大利

项目介绍 PROJECT DESCRIPTION

埃森哲是全球领先的管理咨询、信息技术及外包服务机构，凭借其在各个行业领域积累的丰富经验、广泛能力以及对全球最成功企业的深入研究，发展成为卓越绩效的企业。《财富》全球百强企业中有94家、《财富》全球五百强企业中有四分之三是他们的客户。

此项目涉及三个主要的设计目标：首先是建立起一个操作灵活的区域，顾问们可以通过设置在大堂的一个复杂的电脑系统提前预约办公室。办公室设计的布局非常开放，且满足现代公司的需求：活力、高效，便于信息通讯系统的安装以及工作区的管理。

The project is mainly referable to three design objectives. The first one is to establish an area with great operation flexibility, where the consultants' working location can be booked through a sophisticated computer system located in the lobby. The work environment, for the most part in the open space, recalls the transformation of nowadays to work: dynamism, efficiency, telematic approach extended also to the management of the operative spaces.

第二个目标是设计出五个特别有意思的区域：一楼的前台、招聘室、二楼的品牌展示厅、四楼的团体会议厅和八楼的董事会议厅。这些区域对人——特别是客户和来宾都能产生情绪上的影响。

第三个目标是室内路径的设计。"C" 字形和一个塔形结构的区域分布在建筑的长边上，设计时必须考虑划分员工、顾问、客户和 VIP 客人的不同出入口。

The second objective is to have five areas of particular interest, with a strong emotional impact, especially to clients and guests. The special areas are: reception, recruiting at the ground floor, brand gallery at the first floor, community meeting at the third floor, an area dedicated to the Board at the seventh floor.

The third objective is about the internal paths. The complexity of the building consists of a "C" plan and a tower located on the long side of itself, requiring a careful reasoning about the routes of entry and exit of employees, consultants, clients and VIP guests.

项目必须兼顾室内布局和建筑特点，同时满足客户和地形的要求。设计师将整体空间设想成为一个色彩的容器，一些特别区域用不同的色彩和形式产生视觉冲击。

Progetto CMR has envisioned the building as a colours' container, in particular for the special areas that will become high-impact spaces for colours and forms.

TBWA 广告公司墨西哥城办公室

设计公司：GARDUNO ARQUITECTOS 设计公司、Not Only 建筑事务所
项目地点：墨西哥
项目面积：2781.63 ㎡
摄影师：LAURA COHEN

项目介绍 PROJECT DESCRIPTION

TBWA 腾迈成立于 1970 年，是一家由 Tragos、Bonnange、Wiesendanger、Ajroldi 四个来自不同国家、不同背景、拥有不同经验范畴的广告人合力组成的欧洲广告组织，这在广告公司的创业史上是个特别的先例。腾迈成名靠的是打破常规。作为全球最大的传播集团 Omnicom 的子公司，TBWA 是全球增长最快的跨国广告公司，全球总营业额名列世界第九。

"

设计方案完全打破原有空间结构，建造了一个三层高的中庭。通过中间一层进入电梯，再通过楼梯进入另一层。这样的布局无疑非常冒险，但却与众不同，具有原创性。许多带有数字显示屏的木制小盒子不规则地叠放，垂直贯穿了整个中心区域，每个都可以单独移动，通过图像和声音互动交流，融入用户的思想。

The plan to literally demolish an area, to build a three-level central atrium, entering the elevators through the intermediate level and using stairs to go from one level to the next, was undoubtedly risky, but certainly different and original. An asymmetrical column of superposed wooden boxes—each with independent movement—and a digital screen form an element that travels vertically through the central opening while communicating and blending in the user's state of mind through images and sounds.

Jean-Marie Dru

无处不在的动态和流动元素是每个富有创造力的广告公司的特点，因此设计师在家具和装饰物上安装轮子，如桌子、椅子、文件柜、门和会议室。主会议室的墙壁由可折叠的滑动板构成，这样就可以将图书室和休息室融为一体，创造出一个可供员工聚集的大型会议厅。其余少数墙体由玻璃或白板构成，在上面涂有极具视觉冲击力的标志。自助餐厅向外延伸，构成阳台的一部分，里面摆放着的大沙发可鼓励员工互动交流。

Movements and flows in every possible interpretation are a fixture of creative advertising agencies, which is why we gave wheels to all the furniture and equipment, such as desks, chairs, file cabinets, doors and meeting rooms. The main meeting room is made out of folding and sliding panels, turning into an open area that blends into the library and the main lobby, contributing to create a large meeting space where the agency's staff may come together. The few remaining walls may be modified with the impulsiveness of a marker, since they are made of glass or whiteboard panels. The cafeteria has a large sofa that invites communication and continues outside, as part of the terrace.

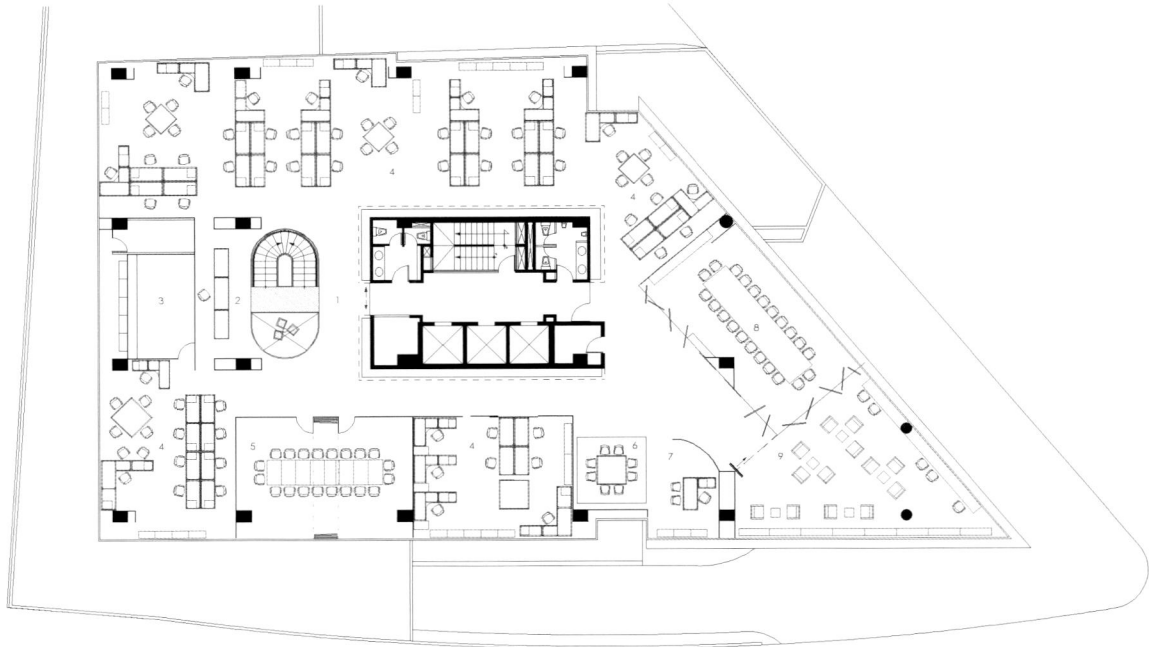

FOURTH FLOOR 四层平面图

FOURTH FLOOR

0 1   5   10 m.

GRAPHIC SCALE

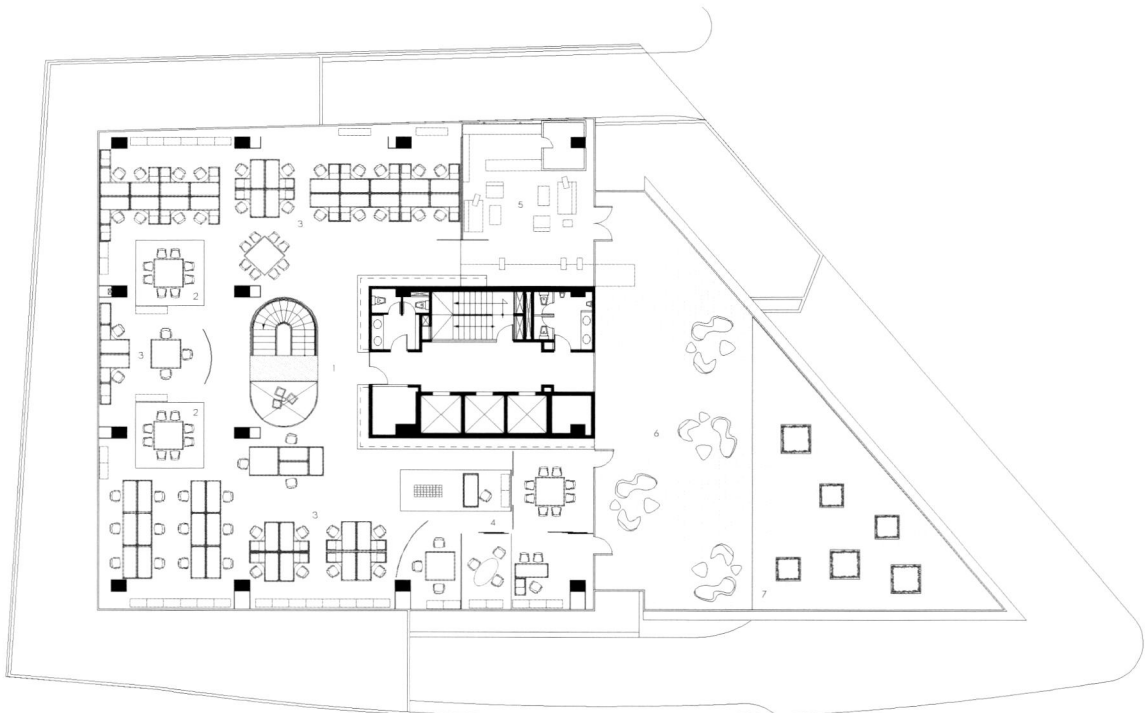

FIFTH FLOOR

FIFTH FLOOR 五层平面图

0 1   5   10 m.

GRAPHIC SCALE

“

整个办公空间的布置富有创新性，营造了一个轻松、以团队协作为中心的环境。

These options are new, fostering a relaxed and teamwork-oriented environment.

图书在版编目(CIP)数据

国际办公新视野：汉英对照 /《国际办公新视野》
编委会主编. —大连：大连理工大学出版社，2011.9
ISBN 978-7-5611-6346-7

Ⅰ. ①国… Ⅱ. ①国… Ⅲ. ①办公室－室内装饰设计
－作品集－世界－现代 Ⅳ. ①TU243

中国版本图书馆CIP数据核字（2011）第139623号

出版发行：大连理工大学出版社
　　　　　（地址：大连市软件园路 80 号　　邮编：116023）
印　　　刷：利丰雅高印刷（深圳）有限公司
幅面尺寸：235mm×310mm
印　　张：23.75
出版时间：2011 年 9 月第 1 版
印刷时间：2011 年 9 月第 1 次印刷
责任编辑：方　　柘
封面设计：王志峰　　李林
责任校对：张媛媛

书　　号：ISBN 978-7-5611-6346-7
定　　价：348.00 元

发　行：0411-84708842
传　真：0411-84701466
E-mail：a_detail@dutp.cn
URL：http://www.dutp.cn